The New Zealand Weather Book

The New Zealand
Weather Book

Erick Brenstrum

CRAIG
POTTON
PUBLISHING

First edition 1998
Reprinted 1999, 2001, 2003

Published by Craig Potton Publishing, Box 555, Nelson, New Zealand

ISBN 0 908802 47 1

Editor: James Brown
Production: Robbie Burton, Tina Delceg, David Chowdhury

Printed by Spectrum Print Ltd, Christchurch

Contents

One of the classic signs of a thunderstorm, these bulges – known as mamatus – typically form on the underside of the anvil of cumulonimbus clouds. They can also form on the underside of high wave clouds formed when the wind blows across a mountain range.

Acknowledgements

First, I would like to thank my colleagues past and present who have taught me so much: in particular the late Tom Steiner who got me started in meteorology and supported my research; Rod Stainer for his encouragement and for giving me the time to write my first articles for *New Zealand Flight Safety*; Ray Smith who gave me the opportunity to write for *New Zealand Geographic*; Alex Neale and the late Henry Hill for the example they set in their studies of the weather; and Augie Auer for his encouragement and support and for being a frequent sounding board for meteorological ideas.

Kennedy Warne at *New Zealand Geographic* has published my articles over the last ten years. His enthusiasm for matters meteorological is unfailing. Some of the explanations and anecdotes used in this book first appeared in *New Zealand Geographic* as did a number of the illustrations. Jim Renwick of NIWA has been my safety net checking the manuscript for errors – any that may remain are entirely my responsibility. Warren Gray and Brett Mullan, also at NIWA, supplied the radar images and SOI graph respectively, and Steve Ready at MetService the cyclone tracks map.

Thanks must also go to MetService for the use of the weather maps, to the Japanese Meteorological Agency (JMA) for the GMS satellite photos, to Wayne Carran for his magnificent avalanche photo, and to Arno Gasteiger for finding just the right 44,000 year old tree to photograph using only my vague directions for guidance. The staff at the Alexander Turnbull Library were so friendly and helpful it makes me want to go back and do another book. Special thanks to my long-time friend Mike Bradstock who encouraged me to embark on this project.

Robbie Burton and Dave Chowdhury at Craig Potton Publishing made it all happen: my thanks to them for their vision and enthusiasm and for dragging me through the hard bits. My thanks also go to Craig Potton for the use of some of his incomparable photos. I owe a special debt to my editor James Brown, who was everything I could want in an editor – enthusiastic, firm, almost invariably right, and gifted in the mother tongue.

Most of all I would like to thank my family, in particular my wife Alex and children Hugh and Anna, for their patience and understanding. This took longer to write than originally intended – most of it borrowed or stolen family time. I could not have got there without your encouragement and support.

Finally, all credit to the weather for turning up.

Erick Brenstrum, September 1998

"Under so plain a title neither abstruse problems nor intricate difficulties should be found. This popular work is intended for many, rather than for few, with an earnest hope of its utility in daily life. The means actually requisite to enable any person of fair abilities and average education to become practically 'weather-wise' are much more readily attainable than has been often supposed. With a barometer, two or three thermometers, some brief instructions, and an attentive observation, not of instruments only, but the sky and atmosphere, one may utilise Meteorology."

Robert FitzRoy, a former New Zealand governor and pioneer meteorologist, in his *The Weather Book*: *a manual of practical meteorology* published in 1863.

CHAPTER 1

Understanding the weather – key principles

The weather at its most violent is a killer. The wind can destroy buildings, sink ships, throw planes from the sky and drive the ocean up onto the land. Heavy rain causes landslides and floods; extremes of heat and cold are a threat to both people and animals.

The weather is also a lifegiver. Rain and sun make plants grow, providing much of the food that sustains us, and a spectacular sunset nourishes the soul. The weather is fascinating in the way it changes, and a good understanding of how it evolves enables us to avoid its greatest dangers and enjoy its benefits.

The air in the sky around us is the factory in which the weather is made. Any attempt to understand the weather must begin by looking at what air is made of and how it behaves. The main raw ingredient in this aerial factory is water, so it is also important to understand how water is transformed as it travels through the sky from gas to liquid or solid.

THE ATMOSPHERE

The atmosphere is a thin layer of gas clinging to the Earth's surface. To get some idea of how thin it is, compare it to an apple cut in half. If the Earth was shrunk to the size of the apple, then the atmosphere would be as thick as the apple skin.

The atmosphere is a mixture of gases, mostly nitrogen (75%) and oxygen (23%), as well as other gases in smaller concentrations. Because the atmosphere is well mixed by the wind, the proportion of these is relatively constant from day to day, with one important exception: water vapour. Water vapour varies from near 0% to 4% of the atmosphere. This is because water is constantly evaporating from the ocean into the atmosphere and then falling out again as rain or snow.

Blanketing the land, a layer of low cloud is trapped under an anticyclone while the mountain tops are in clear skies.

AIR DENSITY

The density of the air is a measure of how much air is present in a cube with sides one metre long. At sea level, air density is about 1.2 kg per cubic metre, while on top of high mountains like the Himalayas, it is only about one third of this. On top of Mount Everest therefore, each lungful of air gives your body only one third as much oxygen as at sea level.

People whose ancestors have lived at high altitude for thousands of years, such as the

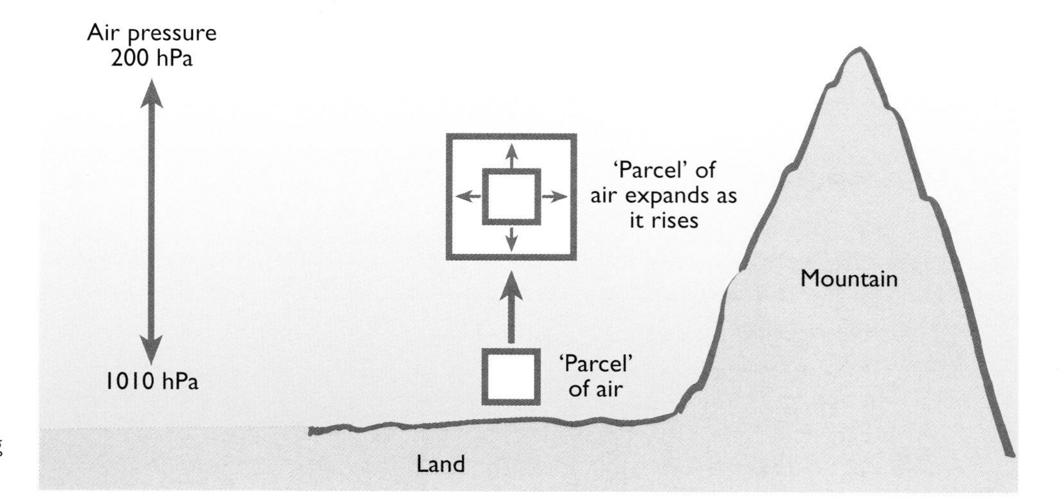

Fig. 1.1. Rising air encounters lower surrounding air pressure and so it expands. This expansion causes the air temperature to fall causing cloud and rain to form.

Quechua Indians of the South American Andes, have evolved larger lungs than coastal peoples. The Quechua also have 50% more haemoglobin in their blood to cope with this lack of oxygen.

Air density is highest at sea level because the force of gravity is always pulling the gas molecules down towards the Earth's surface. But the more molecules gather near sea level, the more frequently they collide with one another and some of these collisions bounce them away from the Earth's surface again. The end result of this is less air the higher you go.

AIR PRESSURE

Air pressure is simply the weight of all the air vertically above a place. The notion that air has weight and is pushing down on us seems a little strange at first. In fact, the weight of all the air inside a family home is approximately the same as the weight of all the people living there.

At mean sea level, the air pressure averages just over 1000 hectoPascals (hPa), while at an altitude of 10,000 m – around the height of Mount Everest – it is nearer 300 hPa. This is simply because there is less atmosphere between the top of Mount Everest and outer space than there is between sea level and outer space. The decrease of air pressure with increasing altitude has a key role in creating the weather.

EXPANDING AIR CAUSES IT TO COOL

When air rises away from the Earth's surface, perhaps in a thunderstorm updraught, it experiences lower surrounding air pressure the higher it goes. This causes the rising air to expand, which in turn causes the air's temperature to fall at a rate of about 1°C for every 100 m it rises.

Other examples of an expanding gas causing a fall in temperature are not hard to find in daily life. For example, air rushing out of a car tyre feels cold on your hand. When you take the top off a bottle of soft drink or beer, the expanding gas in the neck of the bottle cools so fast that a small cloud forms for a few seconds, looking a bit like smoke.

Conversely, when a gas is compressed, its temperature goes up. For example, when you pump up a bicycle tyre you feel the pump becoming hot in your hand; partly because of friction, but mostly because of compression. Similarly, when air descends through the sky it experiences increasing pressure from the surrounding air, and so is compressed and heated.

WATER VAPOUR AND CONDENSATION

Water gas – known as water vapour – is all around us in the atmosphere and we could not survive without it. For one thing, our lungs would begin to dry out and our bodies would dehydrate. The amount of water vapour the air is able to contain depends on its temperature – the warmer the air, the greater the amount of vapour it can hold.

When air contains the maximum amount of vapour possible for the temperature it is at, it is said to be saturated or to have 100% humidity. When saturated air is cooled, some of that water vapour changes back to liquid in a process known as condensation. Sometimes water vapour changes directly to ice without going through a liquid phase. This is known as deposition.

HOW CLOUDS ARE FORMED

If air rises high enough in the sky, expanding and cooling, it will eventually cool enough to reach its saturation point. If the air rises further, and so cools further, some of the water vapour will then condense, forming the tiny liquid droplets that most clouds are made of. Cloud droplets are so small that 200 of them placed side by side would only take up one millimetre.

Cloud droplets stay up in the air because they are extremely light and fall slower than the air around them is rising – thus making no progress towards the ground. The larger a droplet becomes, the faster it falls, until the rising air is insufficient to keep it airborne.

HOW RAIN IS FORMED

As the larger cloud droplets fall, they capture smaller droplets, thereby growing even larger, and falling faster. Eventually, they may grow large enough to fall out of the cloud towards the ground.

This is one of the ways rain can form, and is known as the warm rain process because no ice is involved.

More commonly in New Zealand, rain forms by another process which starts with ice crystals. Ice crystals are more efficient at capturing water vapour from the air than water droplets. Once ice crystals have formed within a cloud, water vapour deposits on the surface of the crystals so fast that the air's humidity falls below 100%. The water droplets then begin to evaporate into the air, providing more water vapour to deposit onto the ice crystals, and so on in a runaway process.

As an ice crystal grows bigger inside a cloud it begins to fall faster. This causes it to bump into water droplets in its path, which then freeze onto it, making it even bigger, so that it falls faster, and so on. It quickly grows many times larger than its original size. Eventually, it falls into warmer air near the Earth's surface, melting into rain before it hits the ground.

LATENT HEAT

One last piece of physics plays a crucial role in our weather, and that is latent heat. In order to make water change from liquid to vapour, heat must be added without the water actually changing temperature. For example, if you heat an uncovered pot of water, its temperature will rise steadily until it boils. Water will then start to evaporate causing the level in the pot to go down, but its temperature will not increase, even as you keep on heating it. The continued heat will instead be transferred to the evaporating water and will be released again by the water vapour when it condenses back into liquid.

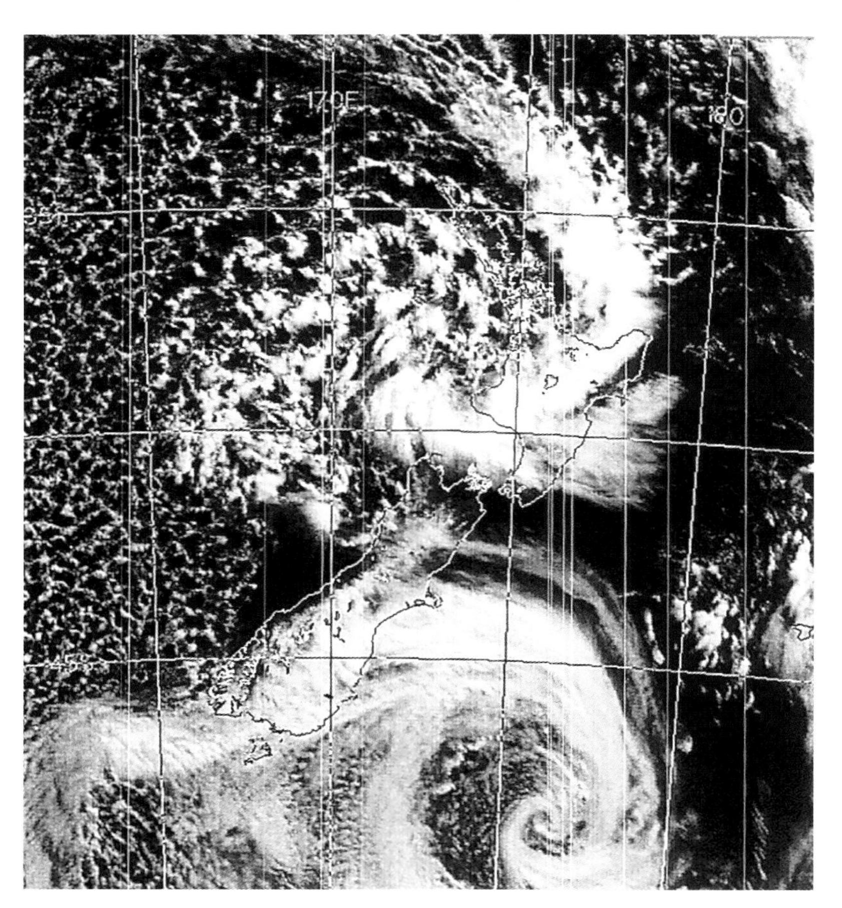

Tall thunderstorm clouds in the cold air behind a front join together in a giant hooked shape over the North Island as a low moves away to the south of New Zealand.

You can experience latent heat for yourself when you get out of the water after a swim. If you do not use a towel to wipe off the drops of water clinging to your skin, they will begin to evaporate, robbing heat from your body as they do so.

Another way to experience this is to have a shower but leave your towel in the hall cupboard. Once the bathroom is nicely steamed up you can turn off the shower without feeling a dramatic cooling. This is because you have your own pet cloud in the room and water vapour is condensing onto your skin at about the same rate that the drops of water on your body are evaporating into the air. However, as soon as you open the bathroom door and let in dry air from the corridor, the equilibrium is lost; the water is able to evaporate off your skin using your body heat, and you feel cold.

The amount of latent heat released by atmospheric processes is vast. For example, one large thunderstorm releases enough latent heat to power a domestic household for 8000 years. Adding this heat to the air can dramatically increase the upward motion that created the cloud in the first place.

TO SUMMARISE

When air rises, it expands, and this makes it cooler. If it cools enough to reach its saturation point, some of its water vapour condenses to form clouds of water droplets. If it keeps on cooling more cloud droplets are produced until rain forms – particularly if the air has cooled enough for ice particles to form.

And so, a search of the weather map to see where rain is likely to fall becomes a search for where the weather map indicates there is going to be rising motion in the atmosphere.

CHAPTER 2

Meet the weather map

The weather maps that appear daily in newspapers and on television show the pattern of air pressure at sea level and are a vital tool for getting to grips with the weather. This chapter will first explain what the elements that appear on the maps are and then show how they relate to the weather by discussing several common situations.

Isobars The numbered lines on the weather map, called isobars, join places where the air pressure is the same, just as contour lines on a tramping map join places of equal height. The numbers are the air pressure measured in hectoPascals (hPa). The lines are drawn at regular intervals of 4 hPa, (although prior to 1995 they were drawn every 5 hPa). Around New Zealand pressure varies from about 1040 hPa to 960 hPa.

Wind From the pattern of the isobars you can work out what the wind is doing at different places. Air moves across the Earth's surface from high to low pressure but is deflected from a straight line path by the effect of the Earth's rotation.

The result is that the wind blows almost parallel to the isobars, crossing them at a small angle towards low pressure. Over the open ocean this angle is about 15 degrees but over land it is 30 degrees or more because of the greater friction over the land than the sea. In the Southern Hemisphere the wind blows clockwise around a low and anticlockwise around a high. In the Northern Hemisphere the flow is reversed.

In general, the closer the isobars the stronger the wind. However, this effect also depends on latitude and whether the lines are curved or not. For example, straight isobars near the tropics at 30 degrees south would cause about twice as much wind as straight isobars with the same spacing at 50 degrees south near Campbell Island. When isobars curve tightly around a low, the wind speed is much less than the spacing of the isobars would suggest. Conversely, when isobars curve around a high the wind speed is greater than the spacing suggests.

A northwest gale sweeps around Solander Island in Foveaux Strait.

A further problem with wind speed concerns the relative temperature of the air and the surface it is flowing across. For example, warm air flowing over cold ocean tends to separate into two layers because the air touching the ocean becomes colder and denser. In extreme cases this can produce a surface wind speed of less than 20 km/h while the speed 100 m up can be as high as 60 km/h and a truer reflection of the isobar spacing.

Lows Areas of low pressure are marked with an L and are referred to as lows or depressions. Lows are associated with rising air. As this air rises it expands and cools, thereby

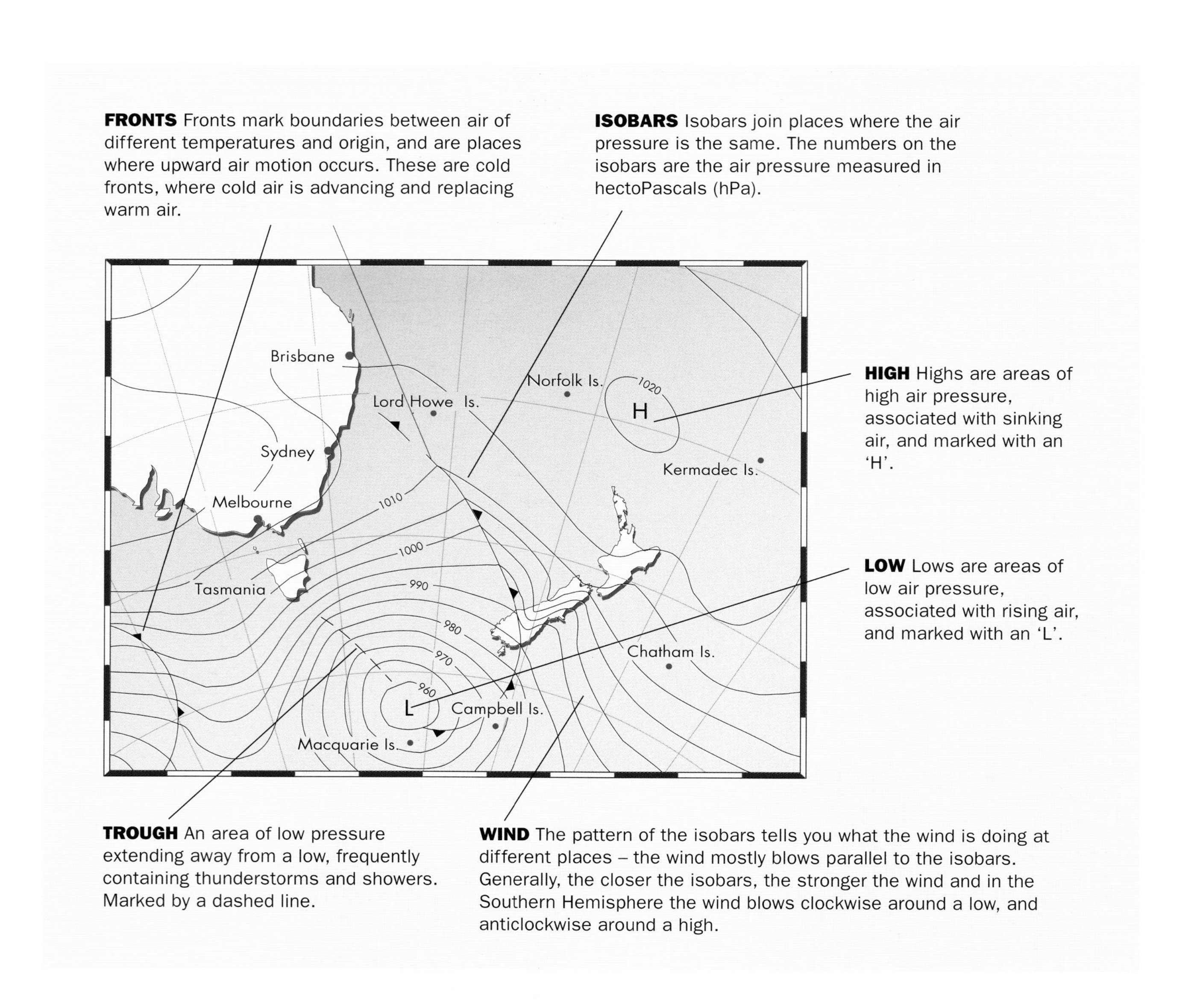

producing cloud and usually rain. However, there can be areas of fine weather near lows because of the sheltering effects of New Zealand's mountains.

Highs Areas of high pressure are marked with an H and are referred to as highs or anticyclones. Highs are associated with sinking air. As it sinks this air is compressed and warmed, thereby causing any clouds to evaporate and so producing fine weather. However, the sinking air does not usually penetrate all the way to sea level. Consequently, low cloud and drizzle are common with highs, especially in winter. Also, in summer, thunderstorms can occur with highs.

Fronts Cutting across the isobars, the lines marked with black triangles or half moons are fronts and are also places where upward air motion occurs. Fronts have broad bands of cloud containing narrower bands of rain, the intensity of which can vary considerably. They occur at the boundaries between air of different temperature and origin, such as air that has come from near the tropics and air that has come from near the South Pole.

Fronts marked with black triangles are cold fronts, where cold air is advancing and replacing warm air. Fronts with black half moons are warm fronts, where warm air is advancing and replacing cold air.

When the triangle and the half moon are side by side and facing the same way, the front is called an occlusion and there is no marked difference in temperature on either side at sea level – although there will be higher up. Occlusions usually develop out of interactions between cold and warm fronts.

The symbols are placed on the side of the front towards which it is moving. When the front slows down and becomes stationary the symbols are marked on alternate sides.

If the line marking the front is dashed it means the front is weak.

Troughs and Ridges A dashed line without triangles or half moons is used to mark a trough. A trough is an area of low pressure extending away from a low and frequently contains thunderstorms or showers.

An area of high pressure extending away from a high centre is known as a ridge, but is usually not marked by a symbol on the map. The effect of a ridge on the weather is similar to that of a high.

WHO GETS THE RAIN? WHO GETS THE FINE WEATHER?

Highs, lows and fronts are all worth more attention and will be discussed in greater detail later in the book. However, the day to day weather is usually produced by combinations of all three rather than any one of them on its own. Moreover, New Zealand is a mountainous country and the major chains of mountains frequently divide the weather into dramatically different conditions on the upwind and downwind sides.

The most common question when people look at tomorrow's weather map is, will it rain where I am? The short answer is that there is a risk of rain if a low or a front is nearby, or if the wind is blowing at strength from the sea onto the land, because rising air occurs with all of these.

Often the most important question is – will the wind blow from the sea towards the land or the other way round? When the strong winds blowing around a depression blow into hills the air is forced to rise over them. The increased upward air motion causes cooling and condensation and can sometimes result in a tenfold increase in the rainfall in the hills compared to down on the coast.

Conversely, once the air has crossed the hills and begins to sink back towards sea level, it rapidly warms by compression and evaporates most of the rain and cloud. If the hills are high enough, this effect can suppress the rain altogether causing an area of fine weather known as a rain shadow to form on the lee side of the hills. The most effective rain shadow in New Zealand is downwind from the Southern Alps.

We will now look at three common weather situations over New Zealand and see where the rain falls.

PREVAILING WESTERLIES

West Coast Flood – Canterbury Nor'wester One of the commonest weather events for New Zealand is a cold front moving over the country from the Tasman Sea with northwest winds ahead of it and southwest winds behind.

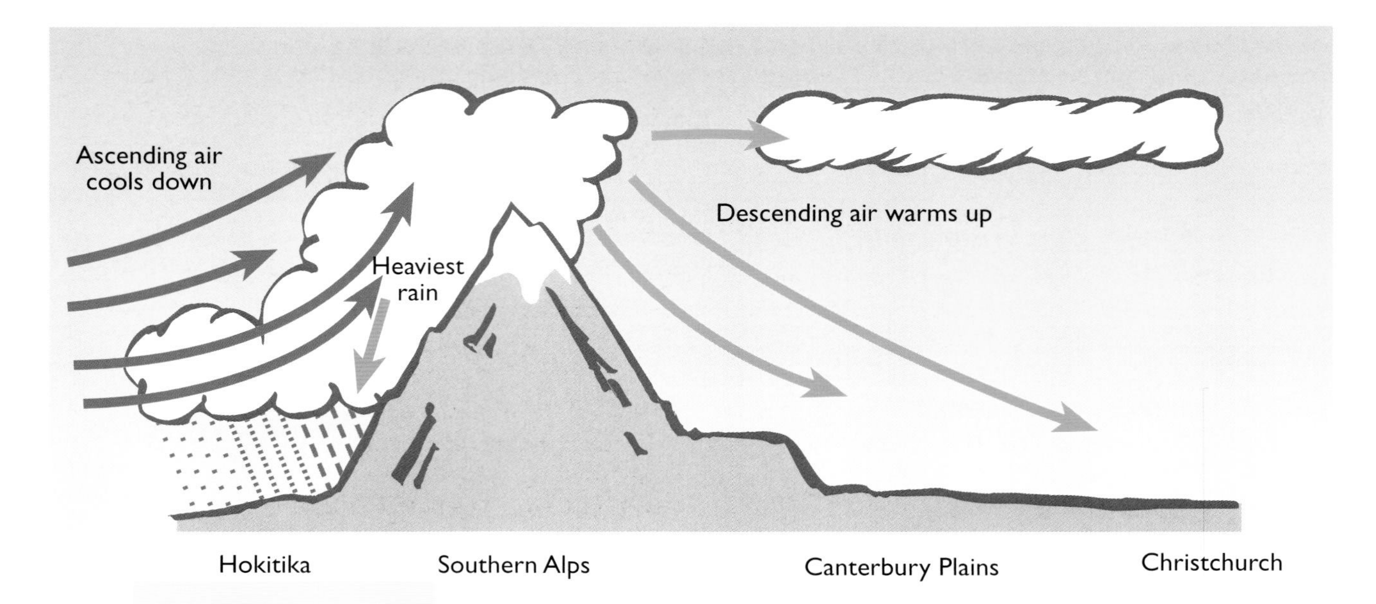

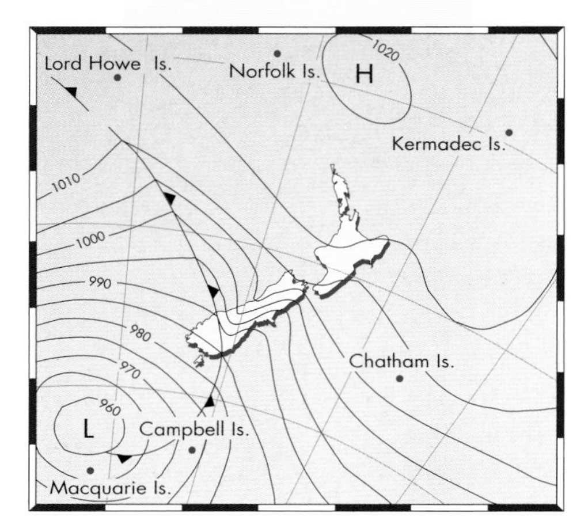

Fig. 2.1. Weather map for 15 October 1988 and diagram of northwesterly over the Southern Alps.

We can see this in the example of 15 October 1988 (fig. 2.1, weather map and diagram). Here a strong northwest air stream is blowing across the Tasman Sea and into the Alps. The air reaching Hokitika is near 100% humidity and only has to rise a few hundred metres in order to cool enough for cloud to form. As it rises further the cooling continues and more and more water vapour condenses causing heavy rain.

The air descends on the Canterbury side of the mountains, warming and quickly evaporating most of the remaining clouds as it does so. When it reaches sea level again its temperature is in the high twenties and about 10°C higher than it was before it reached the mountains. The air has lost more than half its water vapour as rain and is now very dry with a humidity of only about 25%.

The difference in temperature between Christchurch and Hokitika is a measure of how heavy the rain was west of the main divide. This is because all the rainwater that ends up running down the West Coast streams released latent heat as it condensed from water vapour to liquid, making the air much warmer than it would otherwise have been. When this air descended on the Canterbury side of the Alps, it continued warming due to compression and only took a short while to evaporate the remaining water vapour in the clouds. From that moment on, all the heat from continuing compression was available to raise the air temperature on the Canterbury side.

Figure 2.1 shows the isobars ahead of the front close together, indicating strong winds. In Canterbury, the northwest gale is strong enough to blow over a couple of large power pylons and tear roofing iron from buildings. Arcing power lines start a number of fires. Farmers watch in despair as thousands of tons of topsoil, already bone-dry from drought, are blown out to sea. Hundreds of hectares of crops are destroyed, either through windburn or the sandblasting effect of the dust-laden wind shearing off plants at ground level.

This hot dry wind, known as a nor'wester in Christchurch, has dramatic effects on some people. Admissions to psychiatric hospitals increase as do incidents of domestic violence. Winds of this type occur in many parts of the world and are known as föhn winds after the place in the Swiss Alps where their effects on people were first described.

Most föhn winds are not as strong as this example, and not all are so dry as to cause discomfort, particularly downwind of the North Island ranges which are only about half as high as the Southern Alps. In fact, much of the best weather in New Zealand occurs with mild föhn effects in the rain shadows of the ranges.

DEPRESSION OFF THE EAST COAST

Hawkes Bay Flood Uplift of the air by the hills plays a key role in most floods in New Zealand. This is why most floods in western areas occur with winds from between the north and west, and floods in eastern areas occur with easterly winds.

For an example of a flood in the east of New Zealand, look at the weather maps for 1 and 3 June 1963 (fig. 2.2). A depression lies over the north Tasman Sea near Lord Howe Island with a central isobar of 1000 hPa and warm, northerly winds on its eastern side. Meanwhile a front is moving north over the South Island followed by cold, southerly winds. Two days later the low has deepened to have a central isobar of 990 hPa and moved east-southeast to lie near the Coromandel Peninsula. Three isobars cut the North Island coast south of Gisborne, indicating that strong easterly winds are blowing from sea to land over Hawkes Bay causing strong upward air motion there.

In addition, the cold air spreading over the South Island two days before has now become caught up in the circulation around the low. The warm air on the eastern side of the low is now riding up over the colder air because it is less dense. The warm front over the sea east of Napier shows the edge of the warm air at sea level. At higher levels the warm air is already over the land moving into the mountains west of Napier.

The result of all this upward air motion was prolonged heavy rain over Hawkes Bay that caused extensive flooding. Between 100 and 300 mm of rain fell over much of the province in a 24 hour period and one place received around 400 mm. In contrast, only a few millimetres fell in sheltered western areas like Taranaki.

One of the perennial problems in weather forecasting is deciding in which direction

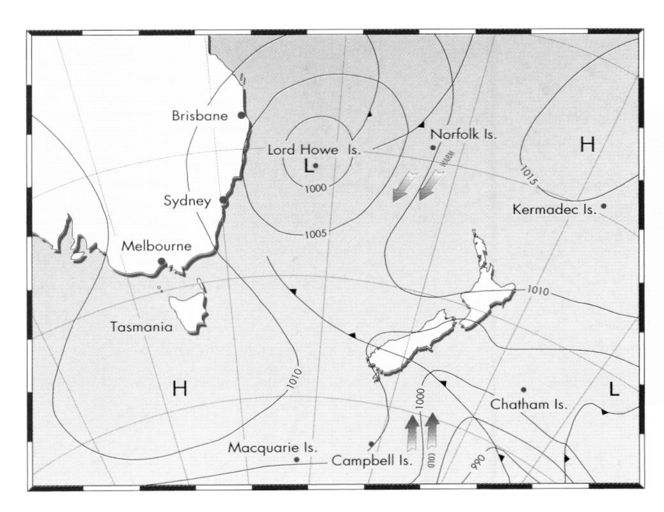

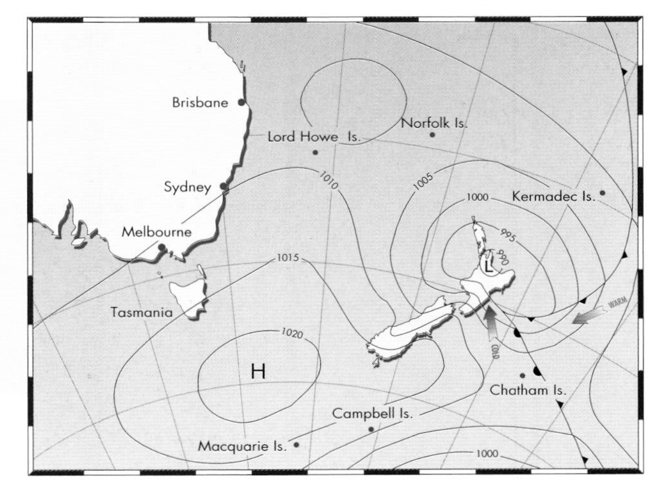

Fig. 2.2. 1 June 1963 3 June 1963

a low will move and how quickly. For example, pretend that, instead of moving east-southeast, the low moves southeast as shown in the second, fictitious, map (fig. 2.3). Now it lies just east of Wairarapa. Consequently, the strong onshore easterly winds affect Kaikoura and north Canterbury, which is where the flood producing rains would occur.

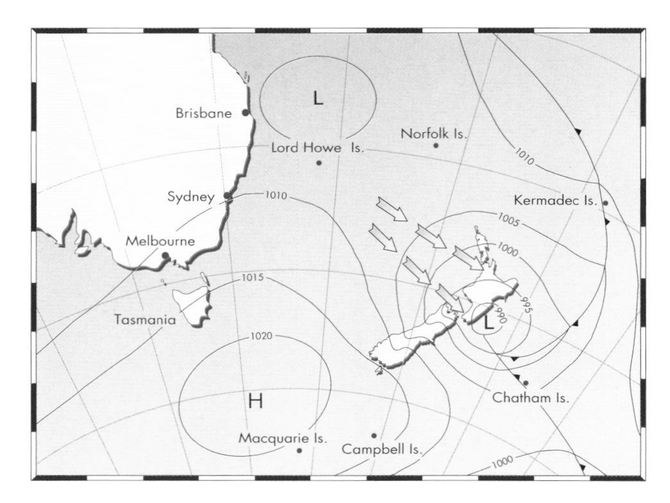

Fig. 2.3. Fictitious map showing how a small change in the low's track radically changes weather distribution. Top line of arrows shows the actual path of the low on 1–3 June.

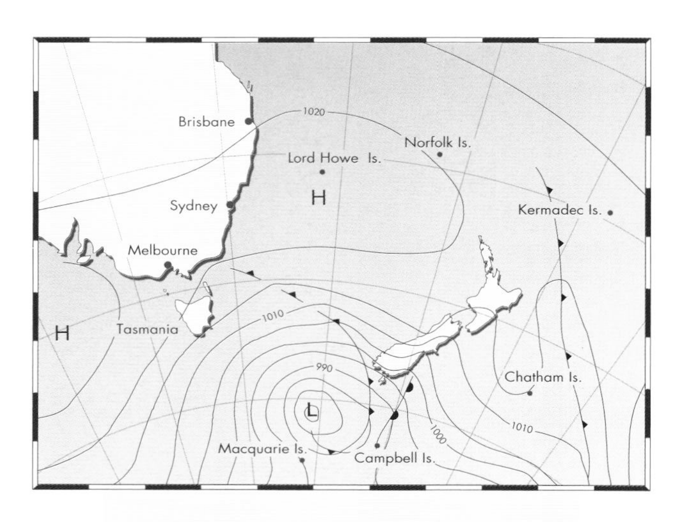

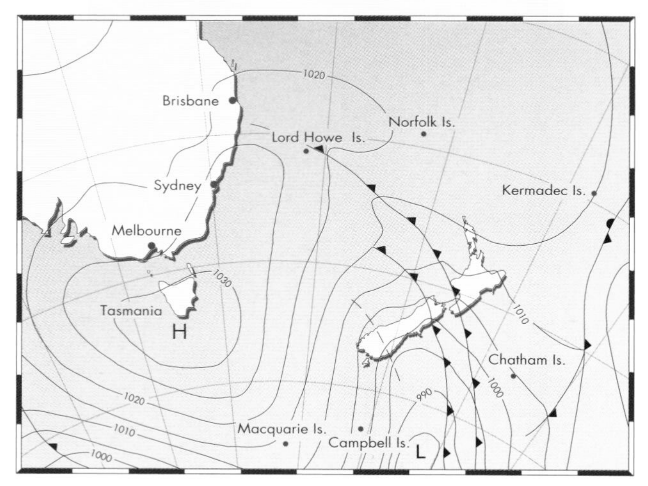

Fig. 2.4. Weather maps for 7 (top) and 8 April 1991.

The difference in the movement of the low centre between this case and the real map for 3 June is only about 20 degrees on the compass, yet the effect would be to shift the flood producing rains 500 km south to Kaikoura.

SOUTHERLY CHANGE

When a front in westerlies sweeps across New Zealand there may be a southerly change spreading showers up the east coasts of both islands, depending on the angle of the isobars behind the front. In the example of 7 and 8 April 1991 (fig. 2.4), the isobars behind the front were initially west-southwest over the South Island and the weather stayed fine in Canterbury because it was sheltered by the Alps.

The map for 8 April, however, shows a trough over Otago followed by southerly isobars. As this trough moved up over the country it was followed by a southerly change which brought showers as far north as Hawkes Bay and Gisborne, while the weather cleared on the West Coast.

What happened next is interesting too. When the southerly change reached the Wairarapa the showers were not particularly heavy or persistent. But a situation soon evolved that caused serious flooding.

On 9 April (fig. 2.5) the pressure at the Chatham Islands was about 1003 hPa and the pressure just east of Gisborne was 1007 hPa. Twenty four hours later the pressure at the Chathams had risen by 12 hPa to 1015 hPa and the pressure just east of Gisborne had fallen by 4 hPa to about 1003 hPa.

This produced a situation where there was a low east of Gisborne with three isobars cutting the North Island east coast indicating strong southeast winds blowing from sea to land. The resulting sustained heavy rain over the Wairarapa caused flooding in a number of eastern valleys. In this case the low was created more by the rise in pressure east of the South Island as the ridge spread over the Chatham Islands, than by the small fall in pressure near the low centre.

Because a large area of hills drained into one valley, the flood rose quickly causing a number of dramatic events. A sheep truck was engulfed by the water surging over a stopbank. The driver survived the night by climbing onto the cab roof. The sheep on the top deck also survived, but all those on the lower deck drowned.

At Alfredton a family was lifted from the roof of their house in the middle of the night by helicopter.

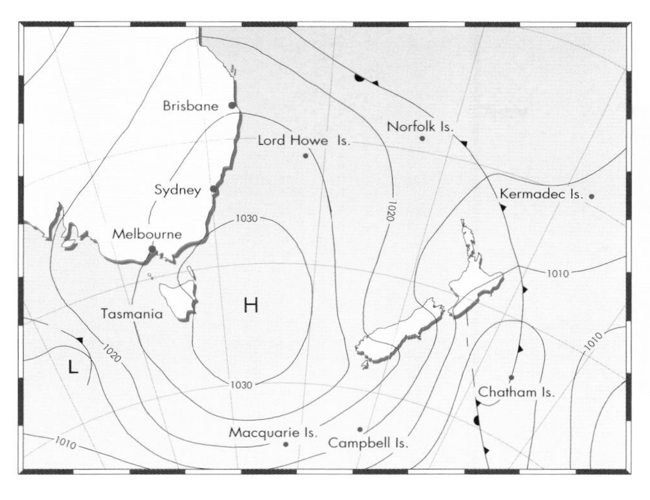

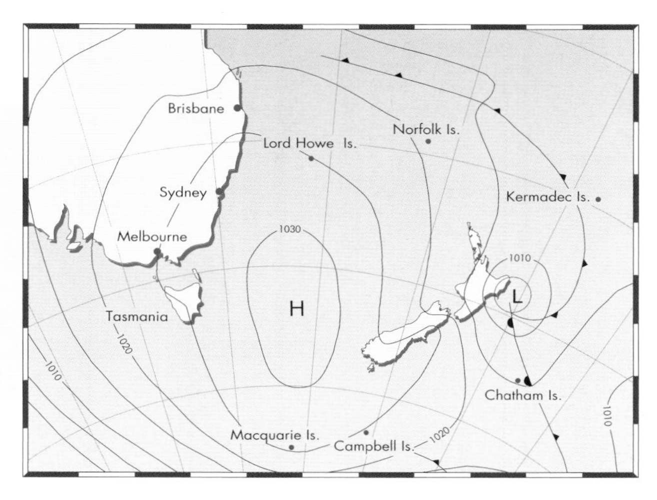

Fig. 2.5. 9 April 1991 10 April 1991

The pilot was guided to the house by neighbours lighting a faint broken path through the darkness with car and truck headlights.

Another farmer went down to his woolshed in the evening to release his horse and dogs so they could reach higher ground. He then swam to his neighbour's holding onto a fence line, in order to release their dogs. When he returned to his property he was trapped in the woolshed, where he spent the night in a bag of dags and crutchings to keep warm.

As well as stock losses many farm buildings were destroyed, roads were blocked by slips and bridges were destroyed. The restoration cost was estimated at $5 million.

The Haast River in South Westland, in high flood.

Overleaf: Northwest arch over Lake Pukaki and Mount Cook, the classic precursor to rain which would fall within the next day or two.

CHAPTER 3

Showers showers almost everywhere

Showers, whether of snow or rain, are usually of short duration – say about 10 minutes – and only affect relatively small areas of about 1 square kilometre at any given moment. They can, however, last over an hour, particularly if they grow into thunderstorms and are anchored over a specific area such as a line of hills.

From a meteorologist's point of view, showers come out of lumpy cauliflower-shaped towers of cloud known as cumulus, whereas rain comes out of flat layer clouds known as nimbostratus. But the two types of clouds can occur together so the distinction between rain and showers is easily blurred. Individual cumulus shower clouds are created by strong upward motion over a small area and usually go through their life cycle in less than an hour. Nimbostratus are associated with weak upward motion over a large area and can last for days.

Shower cloud developing over the Mackenzie Basin in Canterbury.

Because most showers are brief the idea has caught on that a forecast of showers means there will not be much rain. This is not necessarily true. In fact, if a heavy shower stops over one place for an hour or more it can cause a flood.

To understand how showers form we have to remember that through most of the atmosphere, most of the time, the temperature decreases as you go up. This is why there is snow on the mountains.

It is also why pilots in the First World War wore silk scarves. When flying they had to look behind every minute or so to check for enemy aircraft. With so much turning round, the rough military shirts chafed the skin of their necks which then became frostbitten in the subzero temperatures. Silk scarves protected the skin from chafing as well as helping to keep the neck warm.

The upward air motion that produces showers is caused by the air next to the Earth's surface becoming much warmer than the air above. The air at the surface is then less dense than the air above and can rise up through it like a bubble rising through water. As the air rises it expands and therefore cools. So in order to keep rising it must be warm enough to suffer the loss of temperature caused by expansion, and still be warmer than the surrounding air at each new level. Hence we need to modify the old familiar expression 'hot air rises' to become 'hot air rises, only if continues to stay hotter than the air around it'.

When surface air is able to rise freely through the air above because of a strong vertical temperature difference, the air is said to be buoyant and the atmosphere is said to be unstable.

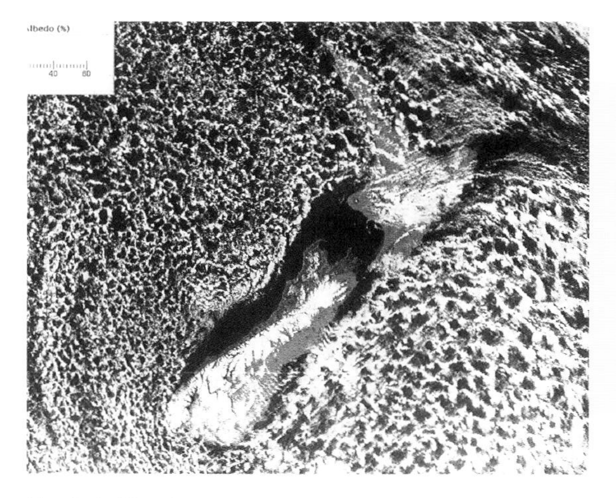

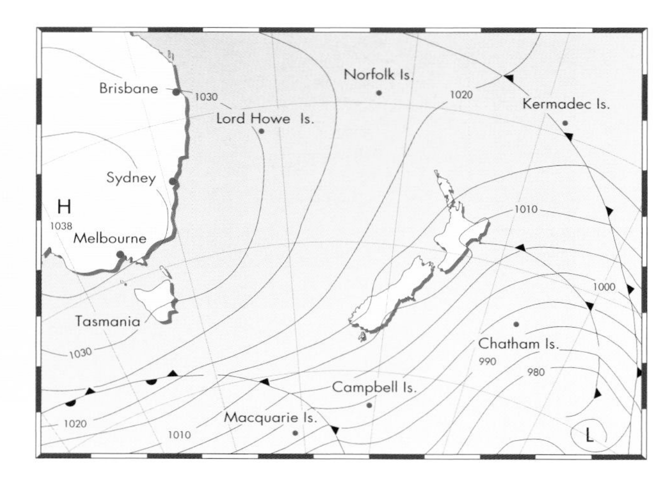

Fig. 3.1. Weather map and satellite photo for 10 August 1994.

WARMING BY THE SEA

The commonest way for air to be heated at the surface is by (relatively) warm sea water. In New Zealand this occurs when a large mass of cold air from the south spreads northwards over the country, usually behind a cold front. In winter, the temperature of this air is typically between 5 and 10°C. So, even though in winter the sea may be too cold to swim in, it is still 4 or 5°C warmer than the southerly air and easily able to transfer heat to it; causing rising, cooling, condensation and showers.

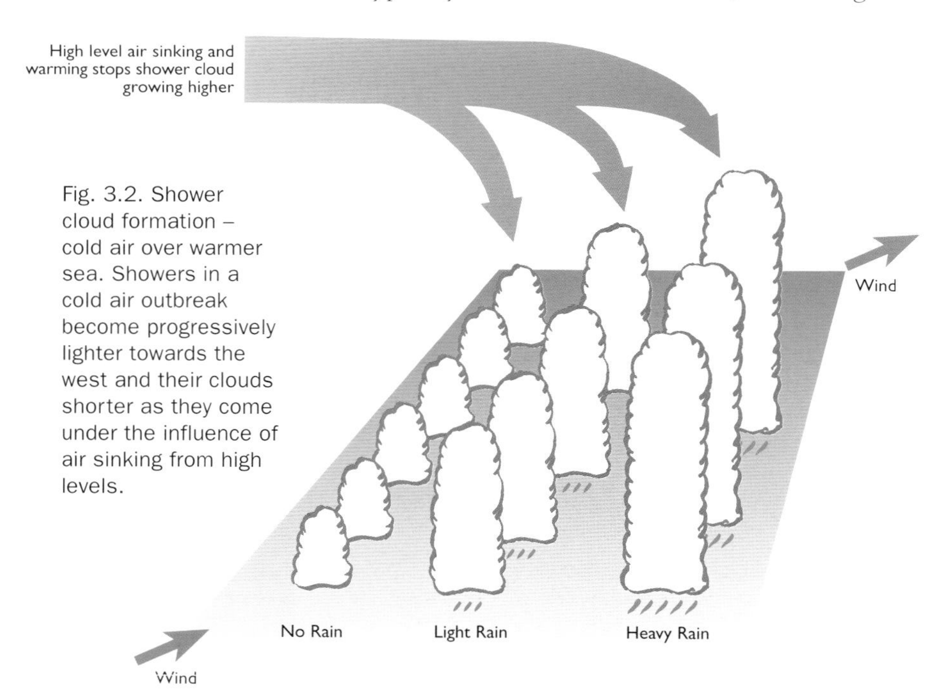

Fig. 3.2. Shower cloud formation – cold air over warmer sea. Showers in a cold air outbreak become progressively lighter towards the west and their clouds shorter as they come under the influence of air sinking from high levels.

A good example of this is seen in the satellite photo and weather map for 10 August 1994 (fig. 3.1). As the photo shows, the showers occur over hundreds of thousands of square kilometres of land and sea at the same time. However, they are not all of the same intensity, nor are the clouds all of the same height. Close to the front east of New Zealand, the clouds are very tall – around 9000 m in height. The showers from these clouds are heavy, with rain falling at a rate of 10 mm an hour or more. Further west, the cumulus shower clouds become progressively shallower and the showers lighter until eventually, when the cumulus are only about 1500 m tall, the showers stop altogether (fig. 3.2).

The gaps between the showers are caused by air sinking towards the surface to replace the bubbles of rising air inside the shower clouds. These gaps are larger next to the tallest clouds, which have the strongest upward air motion and heaviest showers.

The cloud depth at which showers no longer form depends on how cold the air is. Air temperature decreases with height and the level at which it is 0°C is known as the freezing level – although droplets of liquid water don't necessarily freeze there. In fact, liquid water droplets can exist in the atmosphere at temperatures below minus 30°C, although they are not common below minus 20°C. (Liquid water at temperatures below 0°C is known as supercooled water and its presence is essential for the production of hail, as will be seen in the chapter on thunderstorms.)

During winter, very shallow cumulus clouds are able to produce showers because the freezing level is much lower – sometimes as low as 800 m above sea level around Auckland and 200 m above sea level near Invercargill. Generally, the cloud top needs to be somewhat higher than the freezing level for a shower to happen, as ice particles are usually a necessary part of the rainfall process.

When the wind drives cumulus shower clouds into mountain ranges, the extra rising motion as the air is forced up over them causes heavier showers. Once the air has crossed the mountains and begins descending, the air is compressed and warmed and the cloud evaporates. Consequently, areas downwind of the mountains have fine weather and clear skies. The wind on 10 August (fig. 3.1) was southwest – almost parallel with the Southern Alps. So the high country at the bottom of the South Island is blocking the progress of the showers and sheltering the rest of the South Island – except for Banks Peninsula. In the North Island, Taranaki, Wanganui, Manawatu and Wellington are also clear. The white area in the middle of the South Island is snow covered mountain ranges.

The boundary between the showers and clear skies can be clearly seen to the west of New Zealand, but because the top of the North Island curves away to the west, the showers are also affecting Northland, Auckland, and the Waikato.

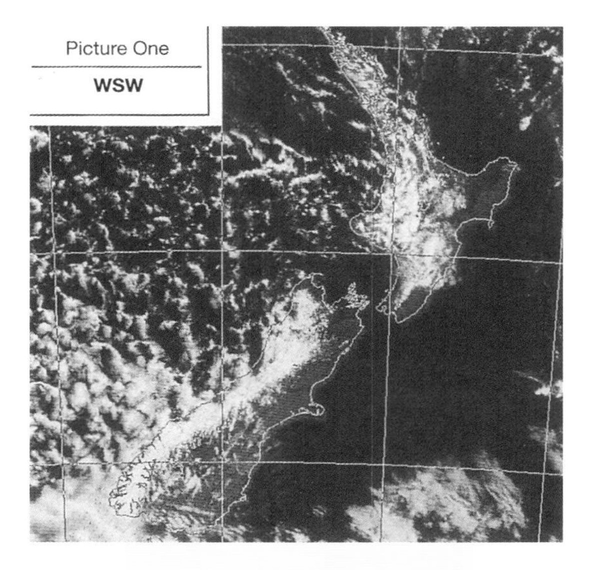

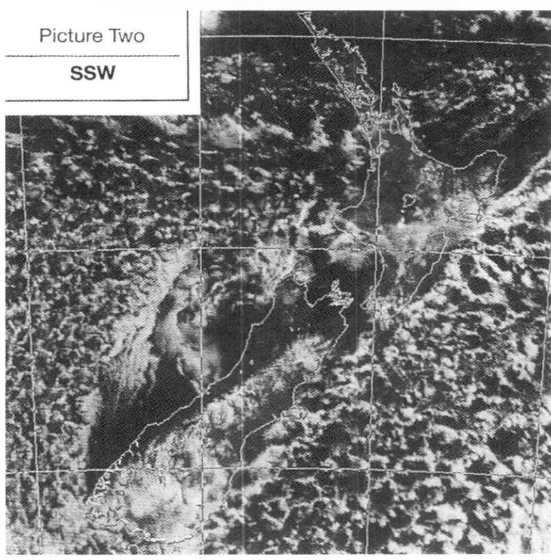

Fig. 3.3.

The position of this boundary between showers and fine weather varies dramatically depending on the wind direction. Sometimes a small wind change is all it takes to swap the showers from one side of the country to the other. Two examples of this can be seen in figure 3.3. In the first, the wind flow over New Zealand is from the west-southwest and the showers affect western areas of both islands. Although mostly blocked by the mountain ranges, some showers have managed to get through the gap between the Tararua and Ruahine ranges near the Manawatu Gorge.

In the second example, the wind flow is south-southwest so showers affect eastern areas. It is fine in the west of the South Island and about to start clearing in the west of the North Island. Because the flow is not all the way round to the south, much of Canterbury is sheltered by the hills of Central Otago and remains clear. Snow showers are falling in Christchurch and Banks Peninsula, while the boundary of fine weather starts over the plains just west of the city. The mountain ranges are again covered in snow.

If the wind direction changed more to the west by about 10 degrees on the compass then Christchurch would clear. If it turned another 20 degrees or so to the south then the showers would spread inland over the Canterbury plains.

In Wellington it was sunny at the time of the photo but there were showers in the Rimutaka Range to the east. A few hours before, the showers had been in Wellington and along the Kaikoura coast, but a minor wind shift took them offshore. Twelve hours later the flow went a little more southerly and the showers came back to Wellington.

Unfortunately, the sheltering we have seen in these examples is not a complete guarantee of fine weather. Showers can develop in clear areas if the heating of the land is strong enough, or if there is strong cooling of the air at higher levels, as we shall see in the rest of this chapter.

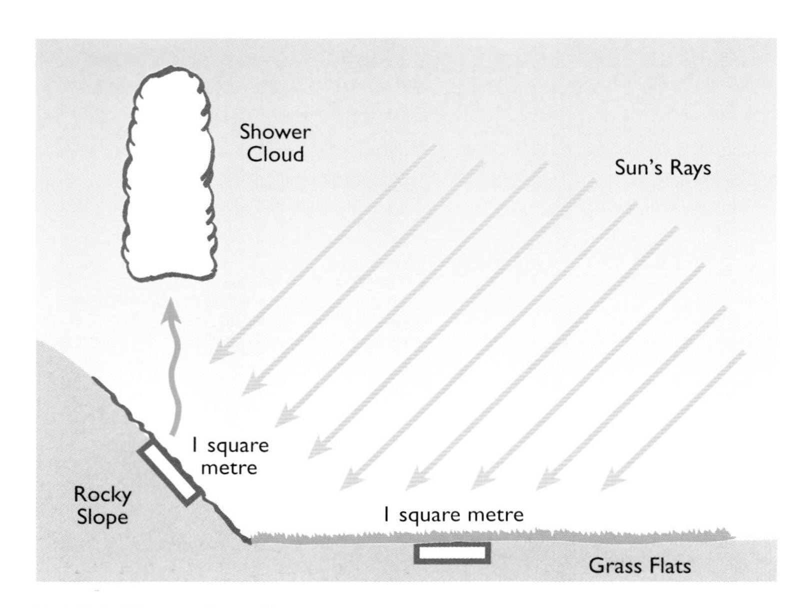

Fig. 3.4. Shower formation – warming by the land. When sunlight heats the land clouds tend to form over the hot spots such as rocky hills. Rocks heat up faster than grass or trees, and some hill slopes are perpendicular to the sunlight and so receive more energy per square metre than flat land.

WARMING BY THE LAND

Aside from cold air moving over warm sea, the other way air is frequently heated at the surface is by land that has been warmed by sunlight. This happens most effectively in the summer half of the year. Showers in these cases are usually heavier than those associated with cold air outbreaks but there are fewer of them and they tend to be isolated. They generally originate over favoured locations and remain there, although sometimes they will drift downwind.

Favoured locations are above places where the land is able to heat up quickly, such as over a ploughed field rather than a forest, or best of all a rocky surface. The slope of the land is also important. A hillside facing the sun and perpendicular to the incoming sunlight will receive more sunlight per square metre than nearby flat land and so will heat up faster (fig. 3.4). The favoured areas will also change throughout the day as the angle of the sun changes.

One of the best places in New Zealand for such surface heating is Central Otago which, paradoxically, is also one of the driest areas. Because of the lack of rain, the vegetation is sparse and the rocky terrain heats up quickly. The showers that develop sometimes grow into thunderstorms with very heavy rain. The result can be flash floods with normally dry creek beds turning to raging torrents in a matter of minutes.

A dramatic example of this occurred in December 1993 when a flash flood brought thousands of tons of boulders, mud and silt down Slaughterhouse Creek near Roxburgh. When the wall of water hit a culvert under State Highway 8, spray shot more than 10 m into the air, causing power lines to arc and cut power to the district.

FIRE

Another way the atmosphere can be heated from below is by fire. If the fire is big enough it can cause the atmosphere to become unstable. The air then accelerates upwards sucking more air in at the base of the fire and causing it to burn more fiercely, so the air rises more rapidly, sucking in more at the base and so on. The surge of air feeding the fire can sometimes cause a whirlwind.

An example of this occurred, with nearly fatal consequences, in January 1974 when the Forest Service started a large burn-off of scrub near Mangakino. Fourteen men were engulfed in a whirlwind as the fire rapidly intensified. They survived by huddling on the floor of a covered-in truck tray for the 15 minutes that the fire roared around them.

The heat was so intense that those touching metal parts of the truck were burnt. One Land Rover was turned into a lump of molten metal and another was thrown from the road by the whirlwind. The hot air rising above the fire caused a tall cloud known as pyrocumulus to develop (fig. 3.5), which then produced a shower.

Fig. 3.5. A dramatic pyrocumulus cloud develops over a controlled forestry burn-off near Nelson.

Pyrocumulus showers will sometimes put out the fires that caused them. This happened several times in the Mohaka State Forest when attempts were made to burn off 1000 hectares of kanuka in fine weather, only to have the fire produce its own shower right overhead.

Volcanic eruptions can also produce showers. Indeed, one of the additional hazards of volcanic eruptions can be giant mud slides, called lahars, set off by torrential rain combining with large amounts of volcanic ash.

COOLING ALOFT

In addition to heating from below, the other important way in which the atmosphere becomes destabilised is through the cooling of air aloft, at heights between 4000 and 8000 m. There are two ways this happens, although both often occur together.

First, winds at high levels are almost always stronger than winds low down, so it is possible for cold air aloft to catch up with warm surface air and move over it creating a strong vertical temperature difference. This then makes it more likely that a bubble of warm air from near the surface will rise freely to high levels and develop a cumulus cloud with a shower.

Second, rather than cold, mid-level air moving over warm surface air, the mid-level air can be cooled by rising. Certain high level wind patterns tend to pile up air in some places – known as convergence aloft – and remove air from other places – known as divergence aloft (fig. 3.6). Under an area of high level divergence, the air at middle levels rises to replace the air being removed. This happens over hundreds of square kilometres and the cooling of the rising air due to expansion leads to colder temperatures at the higher level than existed before. And as with the first cooling aloft scenario, it is then more likely that a bubble of warm air from near the surface will rise freely to high levels

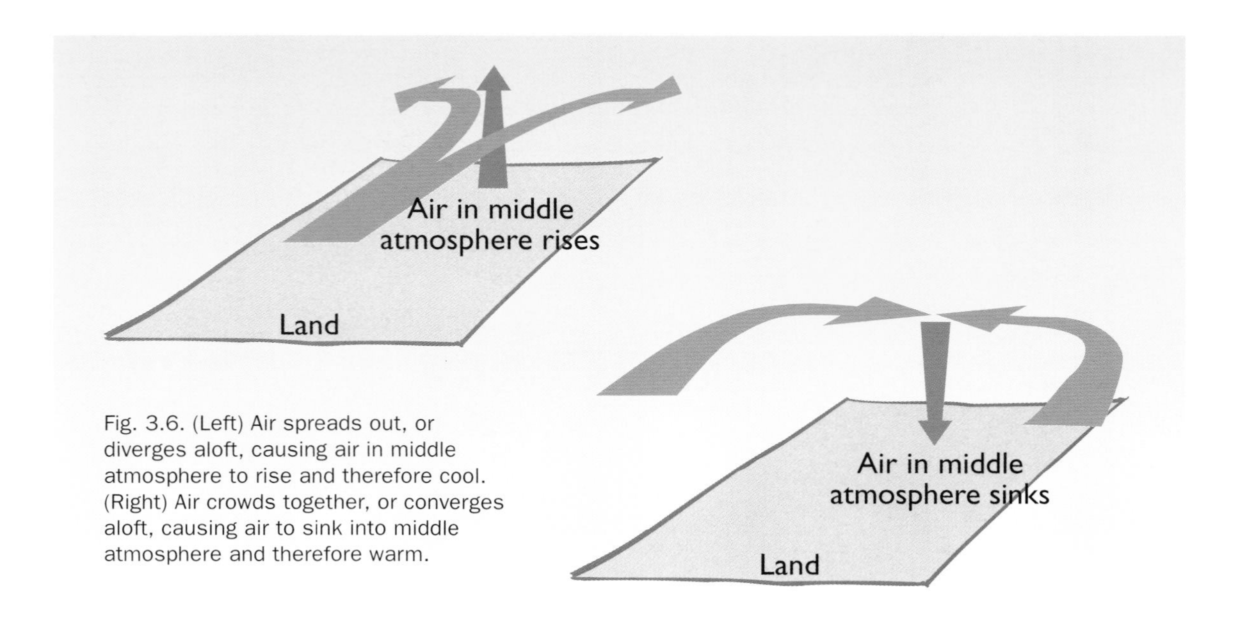

Fig. 3.6. (Left) Air spreads out, or diverges aloft, causing air in middle atmosphere to rise and therefore cool. (Right) Air crowds together, or converges aloft, causing air to sink into middle atmosphere and therefore warm.

and develop a cumulus cloud with a shower.

Cooling aloft caused by divergence is commonplace over New Zealand and plays a leading role in most outbreaks of intense shower activity. Unfortunately, the TV and newspaper weather maps usually provide no clues about this and it is difficult for a person on the ground to spot the signs. So we will now visit the upper atmosphere and examine the three main situations when divergence aloft occurs.

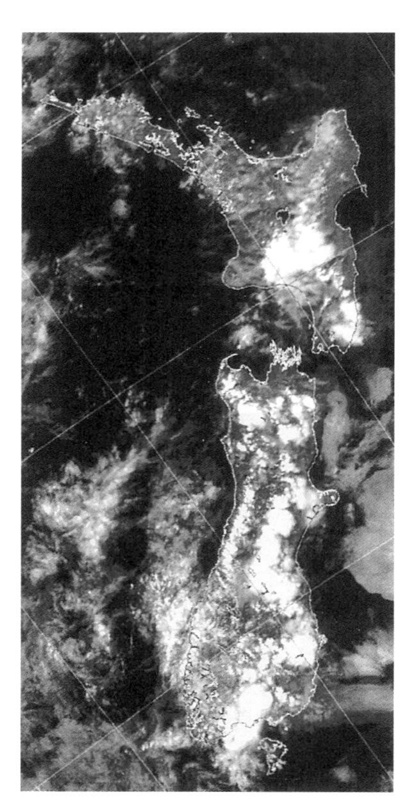

Fig. 3.7. Thunderstorms erupt from Invercargill to Mt Ruapehu on 17 February 1990 – in the middle of an anticyclone (fig. 3.9).

SPLIT IN THE UPPER FLOW

High level cooling occurred over the South Island on Saturday 17 February 1990 helping to cause an outbreak of thunderstorms in the middle of a large anticyclone. The satellite picture (fig. 3.7) taken in the late afternoon shows an almost continuous line of thunderstorm clouds stretching from Southland through Canterbury to the Kaikoura ranges with more in the eastern Wairarapa and near Ruapehu. Flash floods and slips occurred in the Canterbury foothills where rainfalls as high as 28 mm an hour were recorded and a number of high country roads were cut.

The diagram of high level wind flow (fig. 3.8) shows two features that helped produce cooling aloft: a trough to the west of New Zealand and a split of the flow into two separate branches. Cooling aloft occurred downstream of the trough and the split. On this occasion, heating of the land also played a crucial role in the development of the thunderstorms with surface temperatures likely to have reached the high twenties in inland areas.

The showers also developed along the leading edge of a sea breeze spreading in from the coast at low levels. This cooler denser, air would have helped create the upward motion necessary for showers by pushing under the warm air. Once showers began, they would have sucked up more air from the sea breeze. Sea breezes have high humidity so are a good source of water vapour to make rain and also to release latent heat to help increase the upward air motion inside cumulus shower cloud.

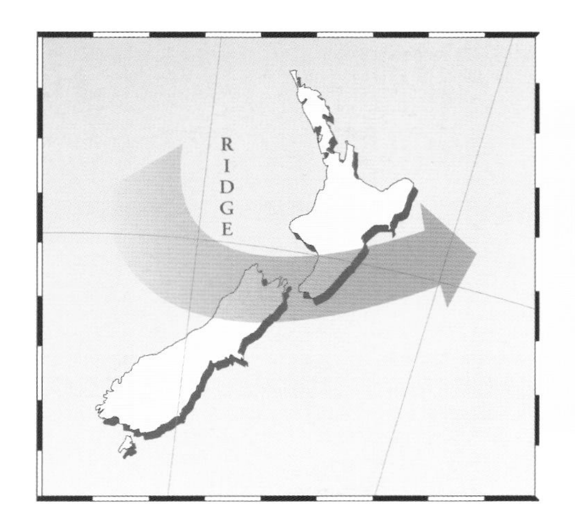

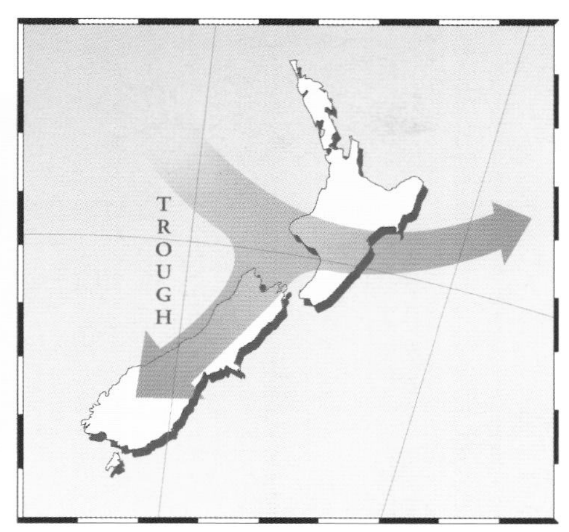

Fig. 3.8. Maps showing winds at about 10,000 m on Friday 16 February 1990 (far left), and the split in high level wind that occurred the following day (left), which led to the thunderstorm outbreak seen in figure 3.7.

The day before this the weather map for mean sea level looked much the same but there were only a few isolated showers. So it is a good example of how a small change to the upper level wind flow can trigger a lot of action without any hint of change in the sea level weather map (fig. 3.9) – a sound reason not to take anticyclones for granted.

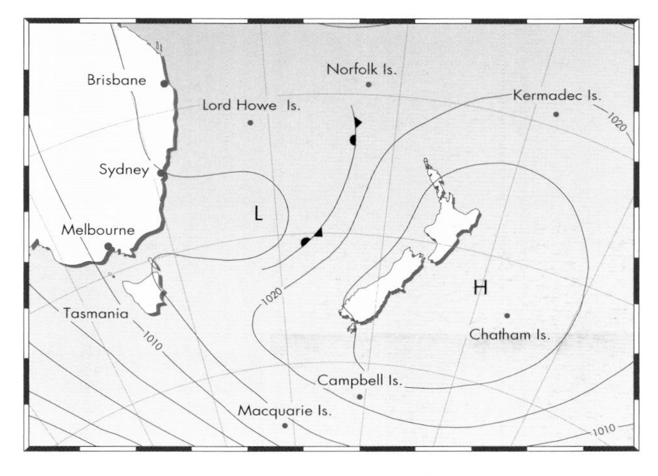

Fig. 3.9. The weather map for 17 February 1990.

INTENSE UPPER TROUGH

The cooling aloft on 17 February was only a couple of degrees. A much larger drop in temperature aloft occurs downstream of an intense upper trough, such as the one that crossed New Zealand on 5 January 1990. The temperature recorded at 6000 m by the weather balloons released at Paraparaumu, just north of Wellington, dropped from minus 15°C to minus 24°C between midday and midnight. There was rain in Wellington with the front, seen on the midnight map (fig. 3.10), but this passed before dawn, then the weather became fine and sunny. Because it was midsummer the heating of the land was at its strongest. This combined with the

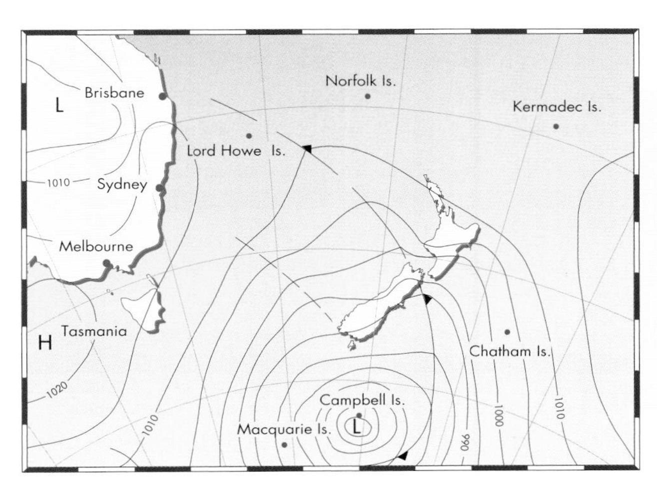

Fig. 3.10. Midnight, 5 January 1990

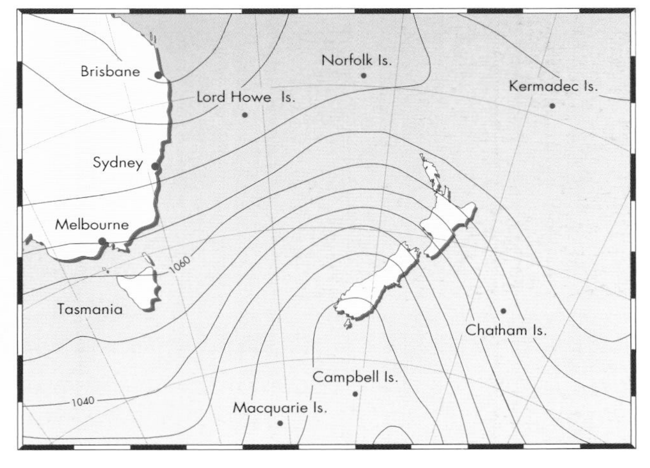

Fig. 3.11. 250 hPa map, 5 January 1990

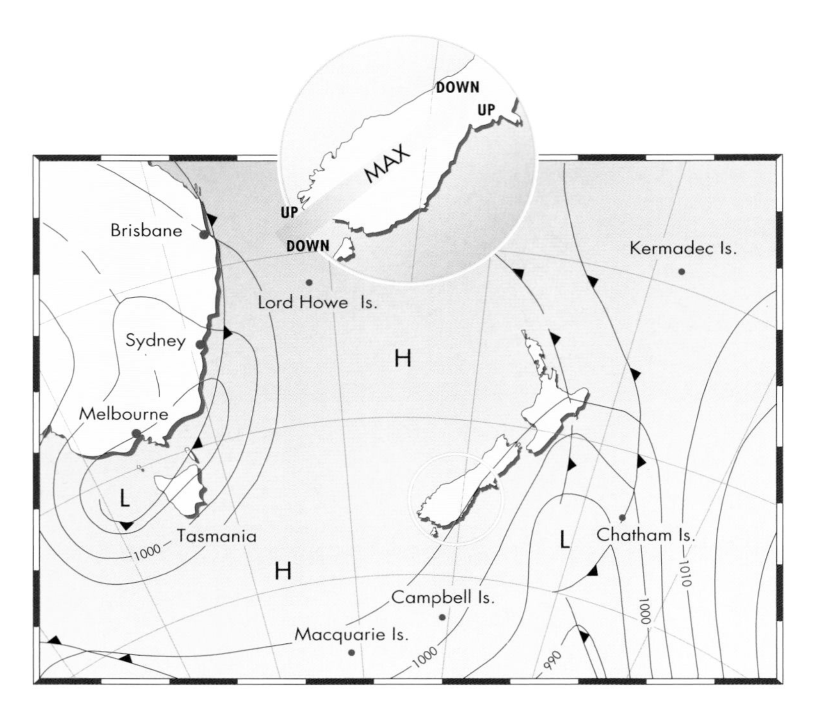

Fig. 3.12. 18 December 1995. Hail showers occurred over Canterbury because of cooling aloft caused by air moving through a maximum in the jet stream. Upward motion ('UP' in the circle) and cooling occur near the polar exit and equatorward entrance to the jet maximum shown in the circle, while downward motion ('DOWN') and warming occur near the polar entrance and equatorward exit.

cooling aloft led to an outbreak of thunderstorms and hail that stopped the national tennis tournament in Wellington for a while.

The intensity of the upper trough can be seen by the number of lines on the 250 hPa map (fig. 3.11, at around 10,000 m), their closeness and the sharpness of their curvature. In these circumstances the surface weather map also shows a similar pattern and can be taken as a guide to a risk of thunderstorms, even in places sheltered from the westerlies, especially if combined with strong heating of the ground.

MAXIMUM IN THE JET STREAM

Winds as high as 180 km/h are commonplace at altitudes of 8000–10,000 m. Frequently they occur in long narrow rivers of wind called jet streams that form along the boundaries between areas of warm and cold air. Jet streams are typically thousands of kilometres long, hundreds of kilometres wide, and just a few kilometres thick. The strongest winds in the jet stream can be as high as 350 km/h and usually occur in a localised area known as the wind maximum. Air moving in the jet stream accelerates into the wind maximum then decelerates as it emerges downwind of the maximum, just like water in a stream accelerates into an area of rapids then decelerates as it emerges downstream.

High level divergence, and therefore upward motion and cooling aloft, occurs in two places near a wind maximum in the jet stream: near the entrance to the maximum on the side of the jet nearest warm air – known as the equatorward entrance – and near the exit to the maximum on the side nearest the cold air – known as the polar exit.

A good example of this occurred on 18 December 1995 (fig. 3.12) when a southwest air stream covered the South Island and showers occurred over inland Canterbury with hail half a centimetre in diameter reported at Woodbury. Weather balloons released from Invercargill measured southwest winds of 160 km/h at an altitude of 10,000 m indicating the presence of a maximum in the jet stream. The polar exit of this jet coincided with the location of the hail showers over Canterbury.

CHAPTER 4

More on highs and lows

Hot and Cold – the big picture

The overall driving force behind the weather is the fact that the tropics receive far more energy from the sun than the poles. This is because the sun's rays enter the tropical atmosphere from almost directly overhead, but as the Earth curves further away from the tropics their angle of entry becomes more oblique (fig. 4.1). Consequently, more of the sun's rays strike a square metre of the Earth's surface in the tropics, than a square metre further towards the poles.

Even in high summer when the north or south poles have sunlight 24 hours a day and more rays reach the ground than in the tropics, the temperatures remain cold because the snow and ice reflect most of this sunlight straight back into space.

The greater heating of the Earth's surface in the tropics accounts for their higher air temperatures. Partly, the air is heated directly by contact with warm land or sea, but more significant heating results from the large amounts of water that evaporate from the ocean. This water vapour travels several kilometres up into the sky before condensing back to liquid cloud droplets and releasing large amounts of latent heat in the process, which warm the air.

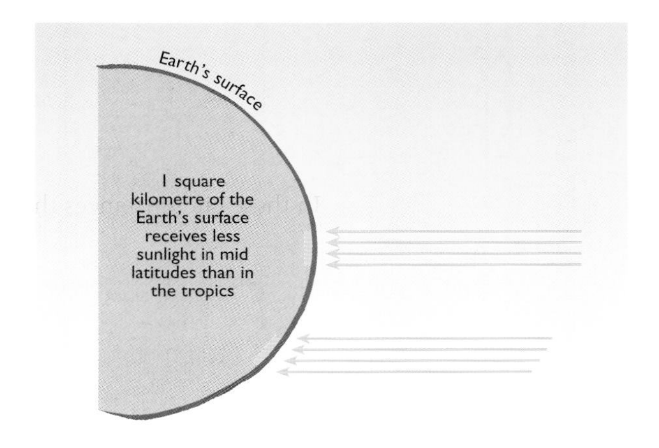

Fig. 4.1. Variation of intensity of sunlight with latitude.

LOWS – WHY DO THEY FORM?

Despite the tropics receiving more heat day by day than polar latitudes, their temperature does not keep inexorably rising. This is because the wind and ocean currents transport heat away towards the poles, and in doing so give rise to the anticyclones and depressions we see on our daily weather maps.

The reason air moves from the tropics towards the poles is to do with pressure. As the lower tropical atmosphere is heated, it expands and lifts the air above it higher into the sky. Consequently, at a height of, say, 6 km above the Earth in the tropics there is much more air above you than there is at the same height over New Zealand. In other words, the air pressure at 6 km above the tropics is a lot higher than the air pressure at 6 km above New Zealand.

Air always moves from places where there is a lot of it – high pressures – to places where there is less – low pressures – in a constant attempt to find equilibrium: thus, air in the tropics flows towards the poles. As it does so, however, its movement is deflected from a direct north–south line to a more westerly direction by the Earth's rotation.

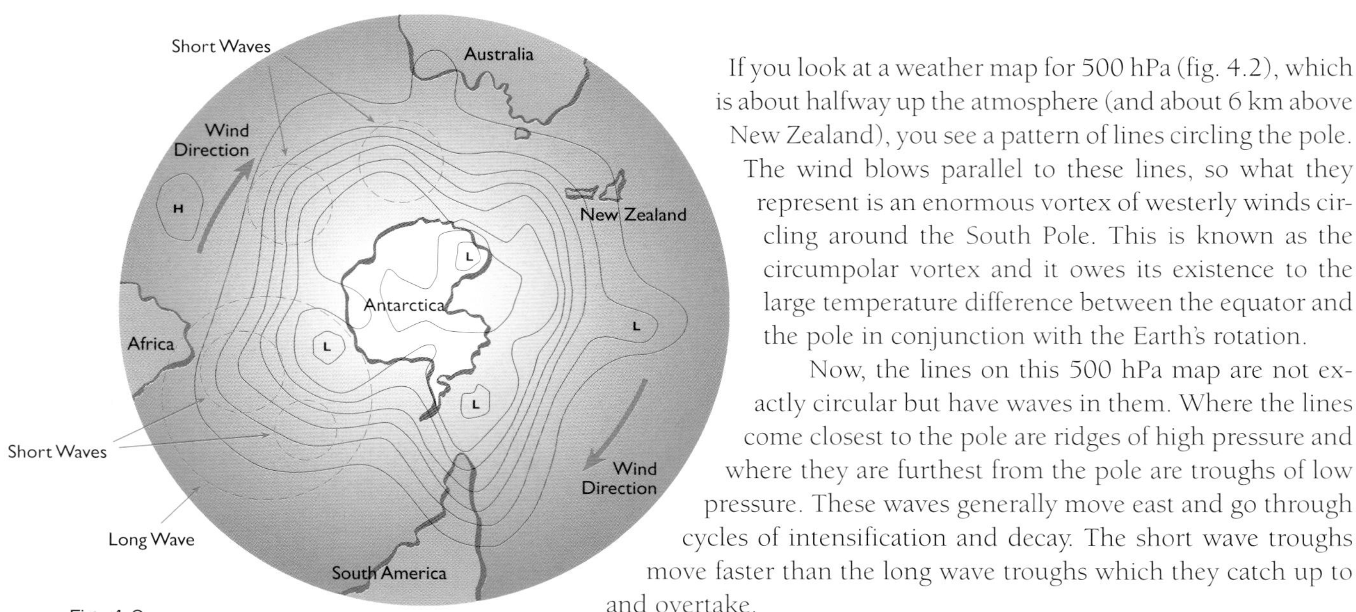

Fig. 4.2.
The circumpolar vortex: this map, depicting weather halfway up the atmosphere, shows waves in the vortex of westerly winds circulating around the South Pole. Short waves move faster than long waves, modifying the weather as they catch up and pass through them.

If you look at a weather map for 500 hPa (fig. 4.2), which is about halfway up the atmosphere (and about 6 km above New Zealand), you see a pattern of lines circling the pole. The wind blows parallel to these lines, so what they represent is an enormous vortex of westerly winds circling around the South Pole. This is known as the circumpolar vortex and it owes its existence to the large temperature difference between the equator and the pole in conjunction with the Earth's rotation.

Now, the lines on this 500 hPa map are not exactly circular but have waves in them. Where the lines come closest to the pole are ridges of high pressure and where they are furthest from the pole are troughs of low pressure. These waves generally move east and go through cycles of intensification and decay. The short wave troughs move faster than the long wave troughs which they catch up to and overtake.

LOWS – HOW THEY FORM

All of this has consequences for the weather, and in particular the development of depressions and anticyclones. As air progresses through the pattern of troughs and ridges, the air in the upper atmosphere tends to be piled up in some places and taken away from others. For example, when air emerges from a trough in the upper atmosphere, it spreads out or diverges, reducing the total amount of air over the area and thus lowering the pressure at sea level. This divergence is greatest when the wind speed is strongest (i.e. the lines are closest together), and where the curve of the lines is sharpest.

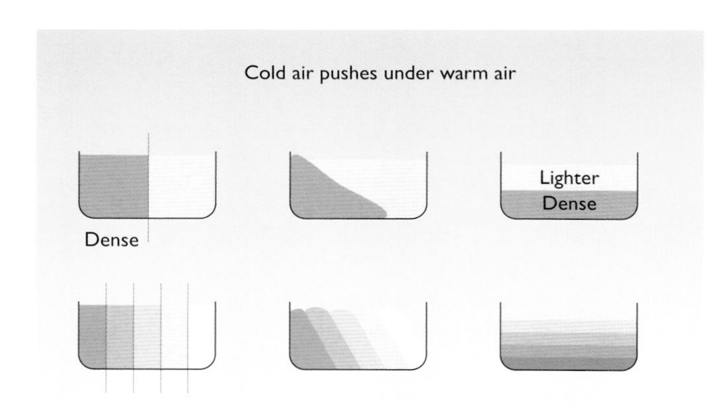

Fig. 4.3. Cold dense air pushing under warm, less-dense air, plays an important role in the development of low pressure systems. Rather than just two layers, however, the atmosphere is more accurately depicted as many layers of slightly different temperature.

Upper level divergence can trigger a depression if it happens over an area where there is warm and cold air in close proximity. As the air pressure at sea level falls, the warm and cold air are drawn together. Because warm air is less dense than cold air, the heavier cold air sinks under the warm air and lifts it (fig. 4.3). The result of this is that the centre of gravity of *all* the air in the area moves closer to the Earth's surface, causing the movement of air around the low to increase as potential energy is converted to kinetic energy.

The conversion of potential energy to kinetic energy as something approaches the Earth can be easily observed by knocking a glass off a bench. As the glass begins to fall it moves very slowly, but its speed increases all the way down to the floor as more and more potential energy changes to kinetic energy.

As the warm and cold air start to circle the low, they cause the upper trough to sharpen, which increases the divergence aloft, which further lowers the air pressure at sea level. This feedback process is why depressions usually have periods of rapid growth early in their lifetime. These typically last 24 hours or less before the upper wind flow evolves to a pattern that no longer favours continued development and the depression is said to have 'matured'.

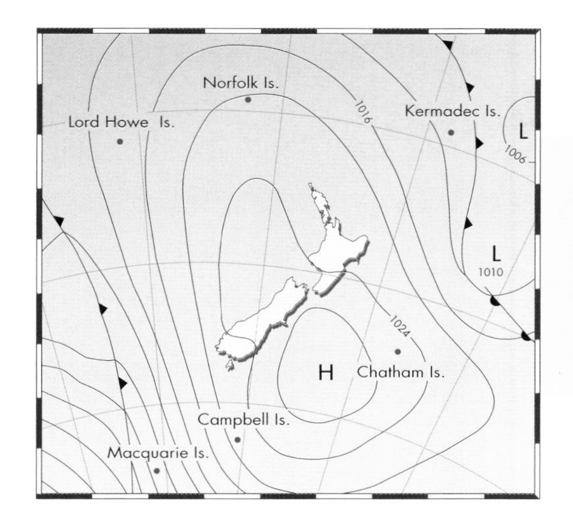

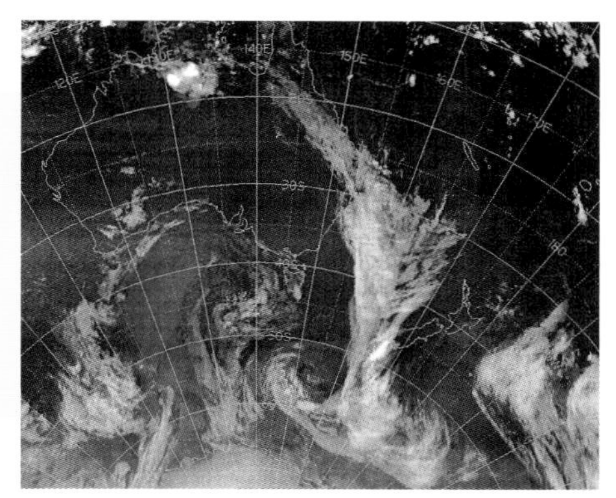

Weather is caused by the unequal heating of the tropics and the poles. Highs and lows are a consequence of moving the warm moist air away from the tropics towards the poles. The dramatic satellite photo shows a giant plume of cloudy air between a cold front and a high (see accompanying weather map) stretching most of the way from the equator to the South Pole, appearing like water being tipped onto the ground. Near Antarctica the low- and mid-level air turns westward and wraps around a low, while high-level air turns eastward and sinks into an anticyclone.

LOWS – WHERE TO FIND THEM

Most depressions are drawn on the weather map with one or more circular isobars around an L, but if the depression is moving quickly, in a westerly for example, the isobar pattern may only show an open wave, as in figure 4.4. Studies of depressions have shown that, particularly in winter, the Tasman Sea is one of the most active areas in the Southern Hemisphere for their formation and intensification.

There are several reasons for this. First, there is a warm ocean current running down the east coast of Australia. Cool air moving over the ocean from the Australian landmass experiences rapid heating and an increase of moisture as sea water evaporates into it. Both of these factors destabilise the air, enhancing the formation of depressions.

Second, the Tasman Sea lies just on the windward side of the Subtropical Jet Stream, which is a belt of very strong westerly winds near 30 degrees south at an altitude of about 10 km. This jet stream is at maximum strength in winter due to the high level outflow of air from the monsoon over the Asian landmass in the Northern Hemisphere summer. The jet tends to slow down as it moves over the Tasman Sea causing air in the upper atmosphere to spread out and diverge, which in turn lowers the air pressure at sea level, and this favours depression development on the poleward side of the jet.

In addition, the Tasman Sea lies downwind of the Great Australian Dividing Range. The passage of westerly winds over these hills can lead to a process known as lee cyclogenesis, which can trigger the formation of a major depression.

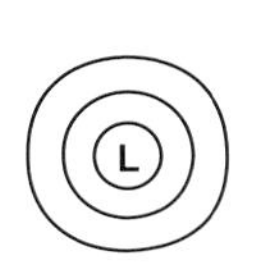

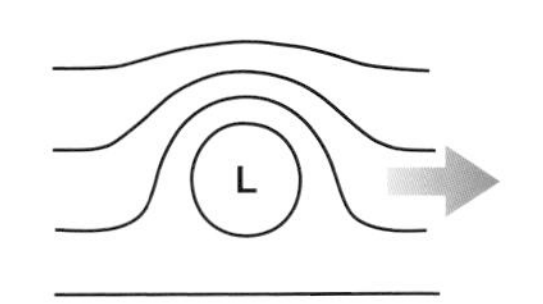

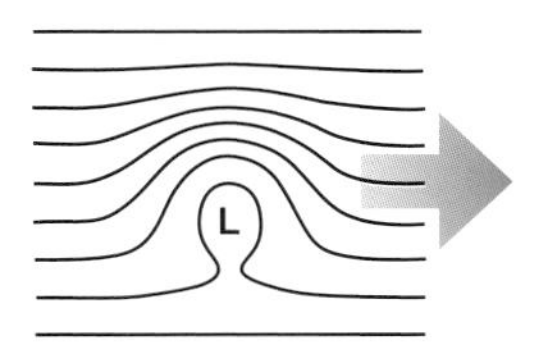

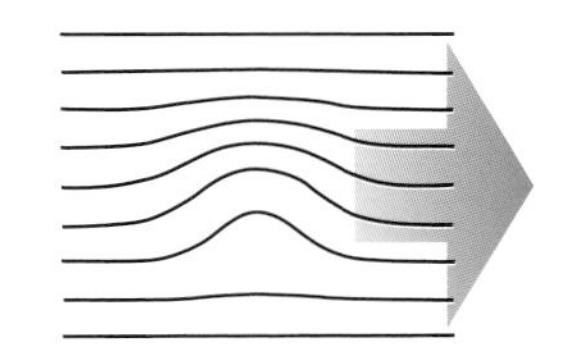

Fig. 4.4.

Most of the depressions that form over the Tasman Sea track east or southeast, passing over or near New Zealand and bringing rain and strong winds to many parts of the country. For example, a low developed over the Tasman Sea on 17 March 1994 then moved east-southeast to lie just west of Buller by midnight on the 18th (fig. 4.5). Warm, moist air had wrapped around the low on its eastern flank and was riding up over cold air

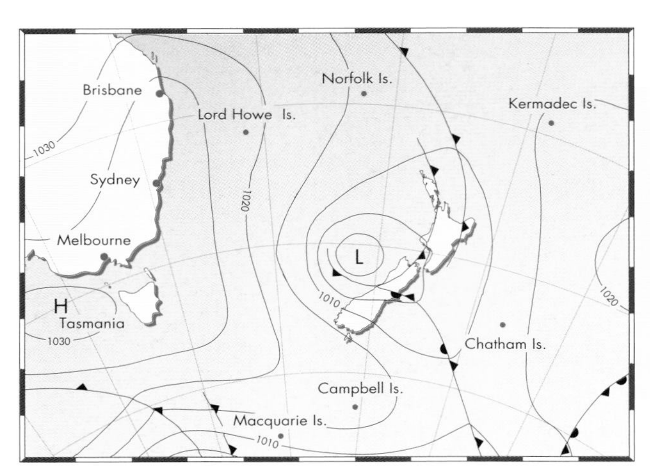

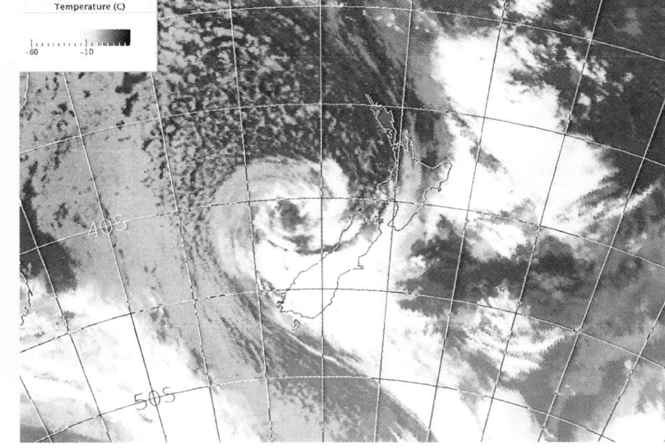

Fig. 4.5. Floods in South Canterbury and Otago. Weather map and satellite photo for 18 March 1994.

as well as the land as it circled into the low centre. The result was prolonged heavy rain over South Canterbury and Otago with widespread flooding. On the 19th, 179 mm of rain fell in 24 hours on the town of Fairlie. The bridge over the Opihi River was destroyed and a $2.5 million timber business on the river bank was wiped out when the river swept through taking tractors, trucks, timber and sawmilling buildings with it (fig. 4.6 photo).

Over 100 mm of rain fell over much of Otago. Roads were washed out or blocked by slips, water entered many homes and one man was drowned when his trail bike was washed away as he tried to cross a swollen stream. Up to 160 residents were evacuated from low-lying areas near the Opihi River and there were serious stock losses in the Hakataramea valley.

Fig. 4.6. At Fairlie, a bridge was swept away and Alpine Timber Ltd's yard was devastated by the flooded Opihi River after 179 mm of rain fell in 24 hours.

LOWS – EXPLOSIVE CYCLOGENESIS

Typically, in a developing depression the pressure will fall about 10 hPa or less in a 24 hour period. A small number, however, develop two or three times as fast and have pressure falls in excess of 20 hPa. These are sometimes called 'bombs' and their development is called 'explosive cyclogenesis'. The extra ingredient firing up the development of a bomb is the release of large amounts of latent heat as water vapour condenses to liquid in the clouds.

The release of latent heat helps to further distort the wind flow pattern in the upper atmosphere in a way that favours upper level divergence. And further upper level divergence then increases the fall of pressure in the low, which in turn increases the upward air motion, which in turn increases the rate of condensation, which in turn leads to more latent heat release, and so on. The importance of feedback processes in the development of weather systems cannot be overemphasised.

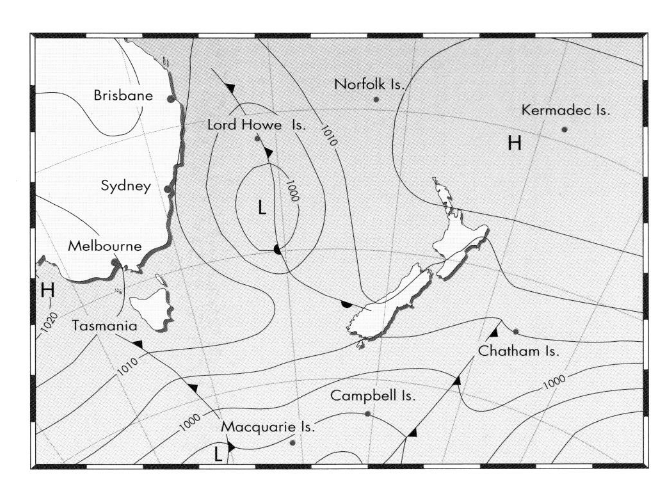

Fig. 4.7. A developing low in the Tasman Sea on 18 January 1988 draws warm moist air from the tropics to fuel explosive cyclogenesis (see fig. 4.8).

Explosive cyclogenesis occurred over the Tasman Sea on 19 January 1988 when a low deepened by over 20 hPa and moved rapidly across New Zealand (fig. 4.7). The air on the low's eastern flank had come from the tropics and so possessed abundant water vapour – which meant plenty of latent heat was available to feed the development once condensation began.

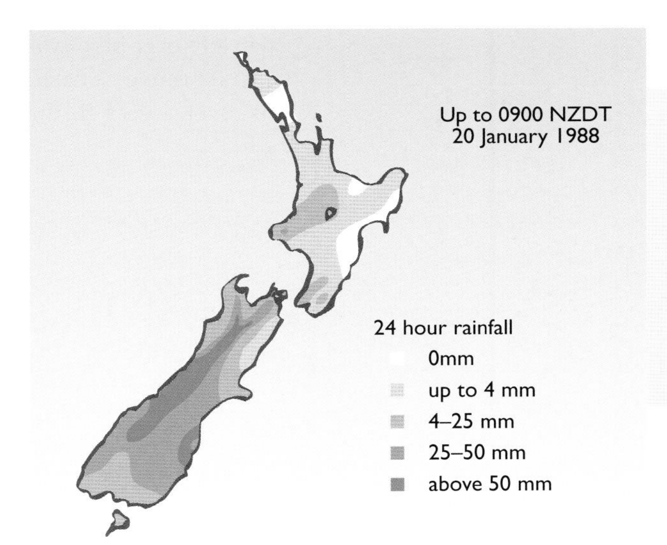

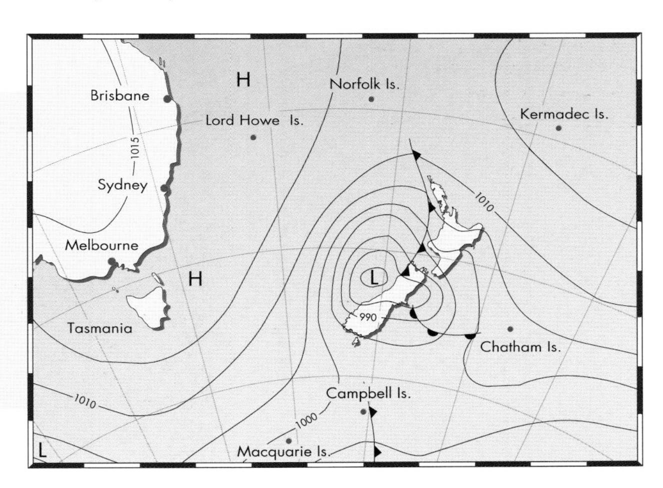

Fig. 4.8. Rainfall statistics and weather map for 19 January 1988 after low in figure 4.7 deepens rapidly and moves onto the South Island.

Heavy rain occurred over parts of the South Island causing minor flooding, but the low was moving swiftly so the rain wasn't prolonged. As it crossed the Alps and emerged just north of Banks Peninsula there was a brief period of southerly gales and a number of houses lost roofs in Lyttelton.

The rainfall map for the 24 hours up to 9.00 a.m. on 20 January (fig. 4.8) shows that the rain in coastal Otago was more than ten times the rainfall near Kaikoura. This is because Otago was subject to strong easterly winds blowing from sea to land on the southern side of the low, whereas Kaikoura experienced northwest winds as the low passed by and so was sheltered from the heavy rain by the mountains.

The heaviest rain in the North Island fell in the central and western high country that was exposed to onshore winds from the west. There was no rain at all in some eastern districts because of sheltering, nor in parts of Northland further from the low centre.

Particularly vicious examples of explosive cyclogenesis occur off the eastern seaboard of the United States during winter when very cold air off the continent meets very warm, humid air over the Gulf Stream. Pressure falls of more than 40 hPa have occurred in 24 hours, creating destructive storms with hurricane force winds that sink ships and bring floods and deep snowfalls on land.

Fig. 4.9. Anticyclones often produce cloudy weather. Although air sinking in an anticyclone warms and dries, causing its clouds to evaporate, the sinking air usually does not reach all the way to the surface, trapping a layer of low cloud close to the ground.

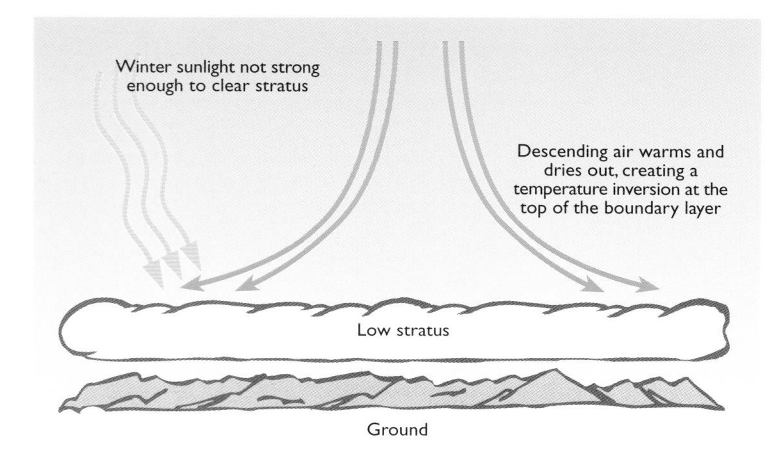

Cloud trapped below anticyclonic inversion on the eastern side of the Main Divide, Hooker Valley, Mount Cook National Park.

HIGHS

Anticyclones are regions of higher atmospheric pressure and are usually just called 'highs'. They are caused by wind patterns in the upper atmosphere that pile up more air in one place – known as convergence aloft – than another. When an anticyclone develops, the air inside it descends from high levels towards the Earth's surface. As it descends, it is compressed – and thereby warmed – by the increasing atmospheric pressure the nearer it gets to the Earth's surface.

Although the air's temperature rises, the amount of water vapour it contains remains the same. But because the amount of water vapour air is capable of containing increases rapidly with temperature, the sinking air becomes, relatively speaking, very dry, and any cloud it may have had, as it began to sink, quickly evaporates.

However, the sinking air does not normally get all the way to the Earth's surface. Instead, it spreads out horizontally around 1500 m above sea level, leaving a layer of colder, denser air trapped between it and the surface (fig. 4.9). The top of the cold layer is marked by a 'temperature inversion'– a narrow layer where the temperature defies the rule by *rising* by several degrees with increasing height. Water vapour and pollutants trapped in the cold air often create a layer of cloud just below the inversion.

In the summer half of the year, this layer of cloud usually 'burns off' over the land during the day because of the strength of the heating by the sun. But in winter, the sunlight, entering the atmosphere at a more oblique angle, is weaker. Then the layer of cloud can last all day and may be thick enough to produce drizzle. So highs do not always bring fine weather!

This is especially so in coastal areas when there is a humid wind blowing from sea to land. In Christchurch, a winter northeasterly can bring two or three days of low cloud and drizzle when an anticyclone is moving slowly over the area. The case from the morning of Tuesday 27 August 1991 to midday on Thursday the 29th (fig. 4.10) was just such

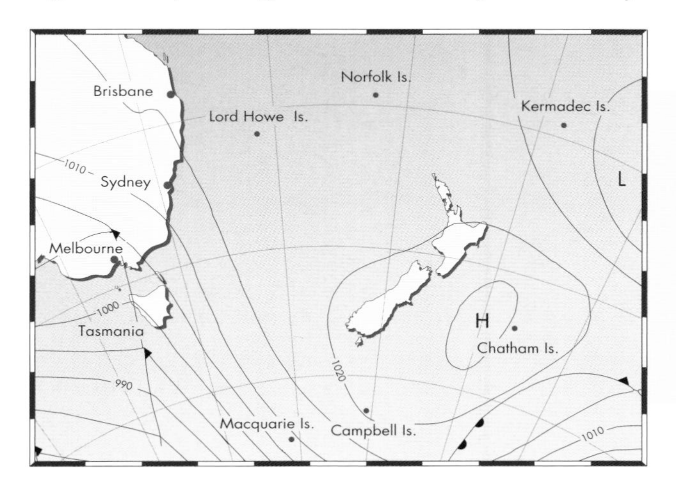

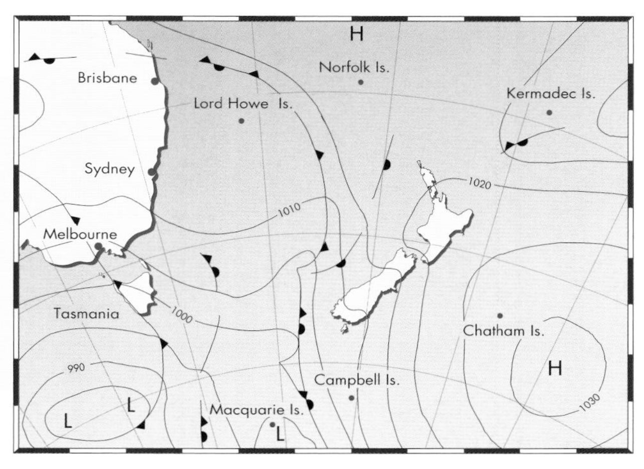

Fig. 4.10. 27 August 1991 29 August 1991

an example. Paradoxically, the weather in Christchurch cleared when the weather map looked more threatening with a front approaching from the west and many isobars spreading over the South Island.

This is because the wind over the Canterbury plains changed direction as the front approached, and blew across the Southern Alps. The air dropped its rain on the West Coast; then, as it descended from the mountain tops, it was compressed and warmed. Because of the wind's strength and turbulence, this dry air penetrated all the way to the surface, clearing the drizzle and low cloud.

Frontal rain often does not reach Christchurch because of sheltering by the Alps. If the wind direction behind the front is between west and southwest, Christchurch will also be sheltered from the post-frontal showers. This was the case in the days following Thursday 29 August. In fact, the next rain in Christchurch came on the Saturday, after all the fronts shown on the map had passed and the next anticyclone developed east of the South Island.

WARM TEMPERATURES

Sometimes the warm air above the inversion does reach all the way to the surface. This can happen when the wind is blowing towards a mountain range. If the inversion is lower than the height of the mountains, the dense, cold surface air cannot rise over them because it is trapped by the warmer, less dense air above it. The only way for the cold surface air to escape is by flowing around the ends of the mountain range.

So in this scenario the air that descends to lower levels after passing over the mountain range is the warm air that was originally at mountain top level when it was upwind of the mountains (fig. 4.11). The low cloud in the colder air remains trapped upwind of the mountains and the descending air is cloud free and warmed by compression as it descends, therefore the weather is sunny beyond the mountains. This allows the land to heat up and warm the air still further.

An example of this occurred on 15 February 1994 in Wanganui and the Manawatu

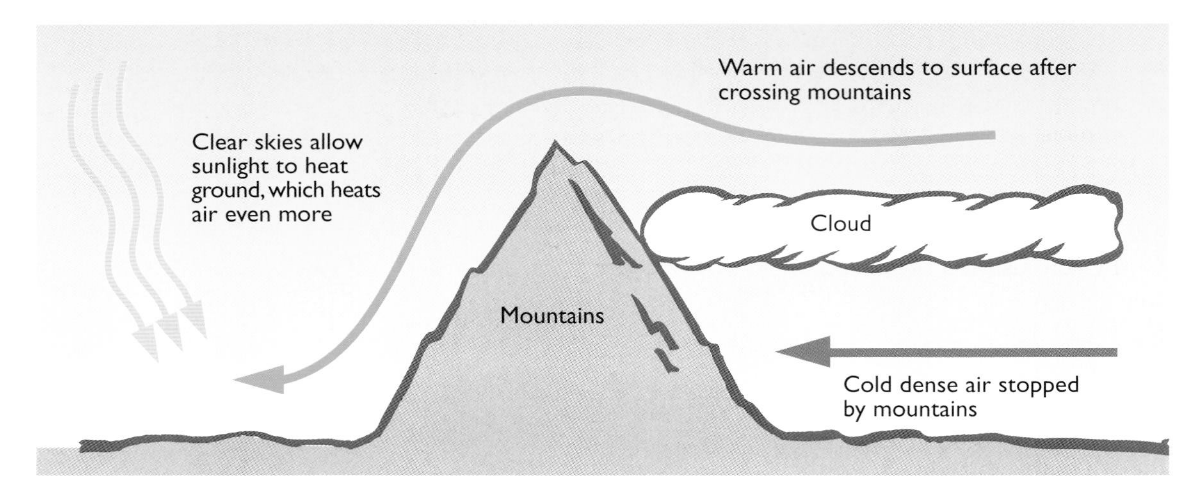

Fig. 4.11. The highest temperatures in New Zealand are caused by a combination of sunlight heating the land plus relatively warm air aloft being further warmed as it descends after crossing ranges of mountains.

Fig. 4.12. Hot weather causes derailment near Wanganui.

when a high was centred over the South Island and an easterly airstream flowed over the Ruahine and Tararua ranges. Railway lines buckled in the heat near Hunterville where air temperatures were estimated to have exceeded 30°C in sheltered places and the temperature of the steel rails is likely to have reached around 45°C. Thirteen freight wagons were derailed, 519 metres of track was damaged and the main trunk line was closed. Freight traffic had to be diverted to the Stratford line, until that was closed by another derailment near Wanganui the following afternoon (fig. 4.12).

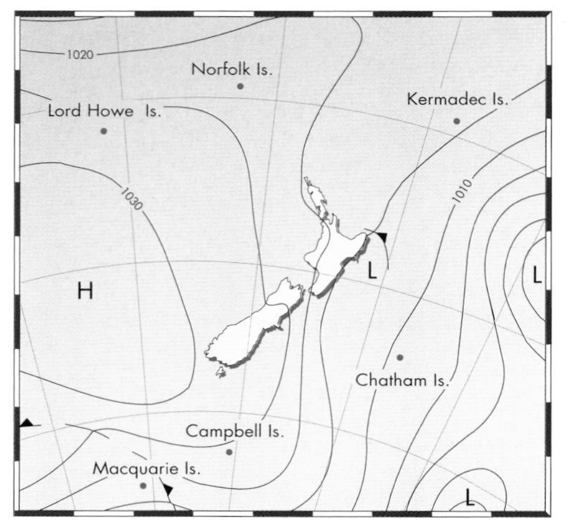

Fig. 4.13. June 1990

BLOCKING HIGHS

One of the important characteristics of anticyclones is the way they sometimes seem to drop an anchor and become stationary or slow moving for a week or more. This is called blocking. It has profound and opposite effects on the weather, depending on whether the high is centred to the east or west of you.

If a blocking high is centred to the west of New Zealand, the country will be subject to prolonged south or southwest winds bringing cold air up from near the Antarctic with showers in those places where the wind blows from sea to land. This happened in June 1990 (fig. 4.13). Showers fell in eastern districts from Canterbury to Gisborne every day for a week, while western and northern areas basked in sunshine.

If a blocking high is centred well east of New Zealand, then the country will experience warm north or northwest winds and periods of heavy rain in those areas exposed to the wind. This was the case during September 1980 when there was a high northeast of New Zealand extending a ridge to Norfolk Island and New

Caledonia (fig. 4.14). A northwest airstream covered New Zealand while a series of fronts moved eastward to the south of Australia before sweeping across New Zealand. Each front was preceded by northwest winds bringing relatively warm moist air from the north Tasman Sea. Western regions had above average rainfall: Hokitika, for example, recorded 385 mm for the month, which was 61% higher than normal.

By contrast, if a blocking high is slow moving over New Zealand, then the weather will be mainly dry, apart from coastal drizzle in winter (fig. 4.10) and some inland thunderstorms in summer (fig. 3.7). In July 1971, for example, a blocking high was slow moving east of the South Island and there was no rain at all in Hokitika from the 5th to the 25th (fig. 4.15). In winter these situations are also likely to cause intense frosts and frequent fog (see Chapter 6).

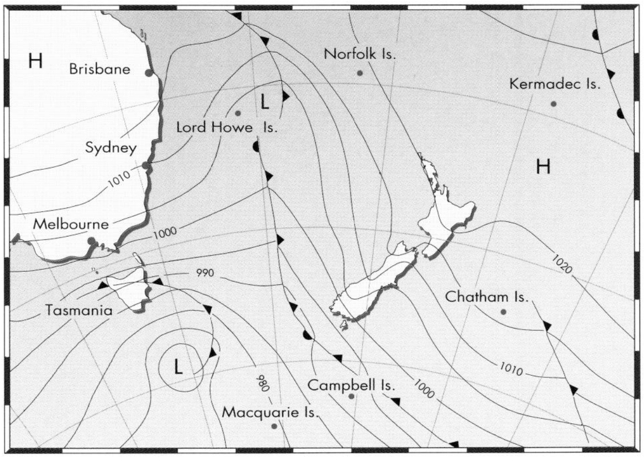

DROUGHT AND FIRE

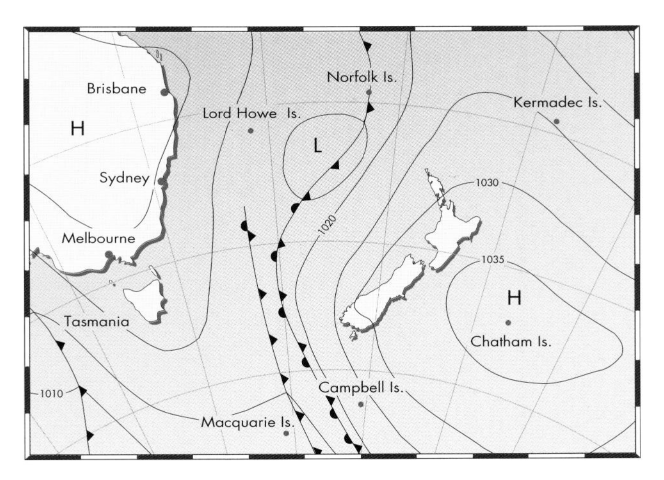

Fig. 4.14 (top). Blocking high during September, 1980.

Fig. 4.15 (below). Blocking high in July 1971.

Prolonged blocking highs combined with persistent winds from one direction (such as those associated with El Niño) so that areas downwind of mountains are always in a rain shadow, are one of the major causes of drought. Their economic effects can be considerable: the Canterbury drought of 1987–88 is estimated to have cost the community $360 million. Aside from the impact on plants and agriculture, droughts are also ideal circumstances for fires.

The likelihood of a fire burning out of control is increased by the length of the dry spell leading up to it. This makes late summer and autumn a time of high risk. Strong winds off the mountains increase the danger in several ways. Due to compression, the air is very dry as it descends from mountain top level, and so dries the vegetation; while the strong winds fan any blaze and their gusty nature can make the movement of a fire erratic and unpredictable.

This was the case on 31 January 1991 when there were fires in eastern districts from Canterbury to Gisborne (fig. 4.16). Buildings were destroyed in a number of places. Sparking power lines, caused by a falling tree, started a fire in a Wairarapa plantation that destroyed 200 hectares of pine trees. A cigarette thrown from a car was the likely cause of a grass fire near Tikokino in Hawkes Bay. People had to be evacuated and hundreds of deer shifted as over 300 hectares of farmland burned. The wind fanning the fires was particularly strong around Wellington where it gusted over 135 km/h. Apart from the standard damage – lines of parked motorcycles toppled, roofs lifted and small planes rolled over – the wind blew confidential briefing papers on the Gulf Crisis out of Parliament Grounds and down Lambton Quay and helped reduce New Zealand's batting to a crumpled heap soon after tea on the first day of the cricket test match with Sri Lanka.

But the worst combination of drought and fire in New Zealand occurred in the summer of 1885–86. This was when large areas of bush were being cleared for farmland in many parts of the country. Trees would be cut down in winter or spring, left to dry, then burnt in summer leaving time for a crop of grass to grow before the following winter.

Rainfall for most of New Zealand had been about 70% of normal for the year to December 1885, but December itself was very dry with only 20 to 30% of normal rainfall. Moreover, there had been many years of incomplete burns preceding this, especially in wetter areas such as Taranaki, so a great deal of suddenly dry timber was on the ground.

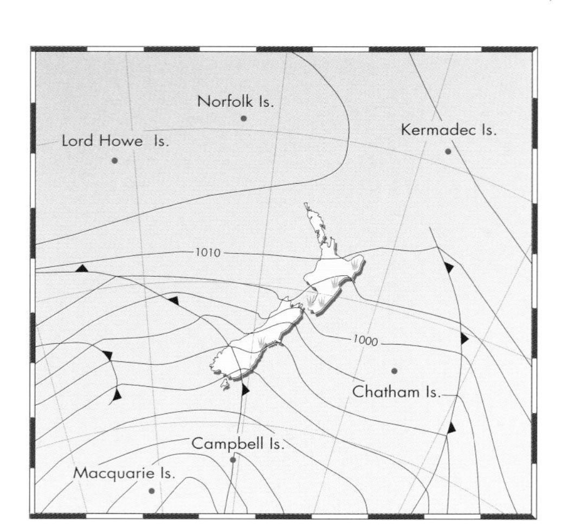

Fig. 4.16. Fire weather – strong hot westerlies lead to fires in the east of both islands.

When the burning season started in early summer, some of the fires got out of control, particularly in areas where strong winds occurred.

In Hawkes Bay, one fire burned continuously from November until the end of January, when it was finally extinguished by heavy rain. The fire burnt back over the same ground two or three times, not only destroying bush, farms and sawmills, but also houses in the towns of Ormondville and Norsewood.

In and around the small Taranaki town of Stratford, a large fire fanned by a southeast gale on 6 January destroyed some 20 homes, many more farm buildings, as well as fences, crops and pastures. Remarkably, no one was killed, although many people suffered burns and temporary blindness after fighting to save buildings, and many animals perished.

There were some very narrow escapes. When fire swept through the bush clearing where Lawrence Woodruffe was farming, he had just enough time to rush his wife and baby to safety in a nearby creek bed before attempting unsuccessfully to save the farm buildings. As flames engulfed his house, he found his retreat blocked and sought refuge down his well. The windlass support quickly burnt through, sending the heavy wooden drum crashing in on him. Woodruffe had to flatten himself against the side of the well and then douse the burning timber. When he climbed out, he found his horse burnt to death only metres away.

At a nearby farm, after failing to save their house, the family members lay down in the potato patch, covering themselves with green leaves. They survived, although their clothes were scorched on their backs. Two pigs which had fled with them were roasted to death.

Bushfires have long been a hazard in Australia as shown in this etching from 1879. In New Zealand bushfires were also a major hazard when forest was being cleared for farmland. During the drought of 1885–86 one such fire engulfed the Newmans coach on the road from Nelson to Murchison. Unable to turn on the narrow road the driver was forced to continue through the fire as the flames singed the horses and blistered the coach paintwork until they reached the safety of a creek bed.

CHAPTER 5

A closer look at fronts

Fronts are bands of cloud and rain, typically thousands of kilometres long but only tens of kilometres wide, that form at the boundaries between areas of warm and cold air. They are usually accompanied by a wind change, a change in temperature and a rise in pressure. The wind change can be seen on a weather map in the way the isobars abruptly change direction as they cross a front.

The traditional explanation for the formation of cloud and rain with fronts is that because cold air is denser than warm air, when the two come together, the warm air is forced to rise. In the case of a cold front, the cold air burrows under the warm air, pushing it up; in the case of a warm front, the warm air rides up over the cold air. Either way, the rising air is cooled by expansion causing water vapour to condense and form clouds and rain (fig. 5.1).

But this explanation is now acknowledged to be only part of the story. In fact, a lot of cloud is formed ahead of the front by other mechanisms. Studies of satellite and radar images, and flights into storms by specially equipped research aircraft, have given rise to the idea that cold fronts are preceded by 'warm conveyor belts' of air. These are thousands of kilometres long and start close to the Earth's surface in or near the tropics before sloping upwards as they move towards the poles. Altitudes of 6 km are reached before they turn away from the front and start descending again (fig. 5.2).

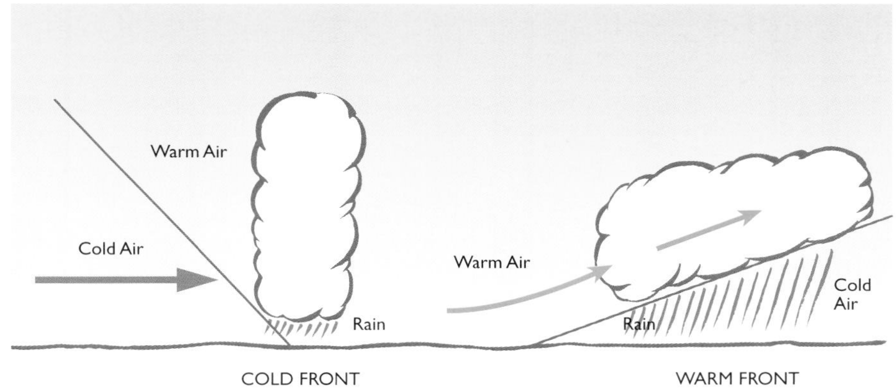

Fig. 5.1. The familiar explanation for frontal weather, first worked out in the 1920s by Norwegian meteorologists, emphasised air movement perpendicular to the front as the cause of upward motion. With a cold front, cold air drives under warm air forcing it to rise; with a warm front the warm air rides up over the denser cold air.

Studies by aircraft also show that the temperature difference across a cold front is concentrated in the lowest couple of kilometres of the atmosphere, rather than extending to great heights as the traditional explanation had it. This is because the warm air ahead of the front is rising and therefore cooling at the same time as the cold air behind the front is sinking and therefore warming. Both of these processes reduce the temperature difference across the front in the middle atmosphere.

Most of the action with fronts over New Zealand is caused by cold fronts and the warm conveyor belts of air ahead of them. Warm fronts are usually weak over the country – although they can be strong once they penetrate further south towards the subantarctic islands.

Fronts are created by wind patterns that blow warm air and cold air towards each

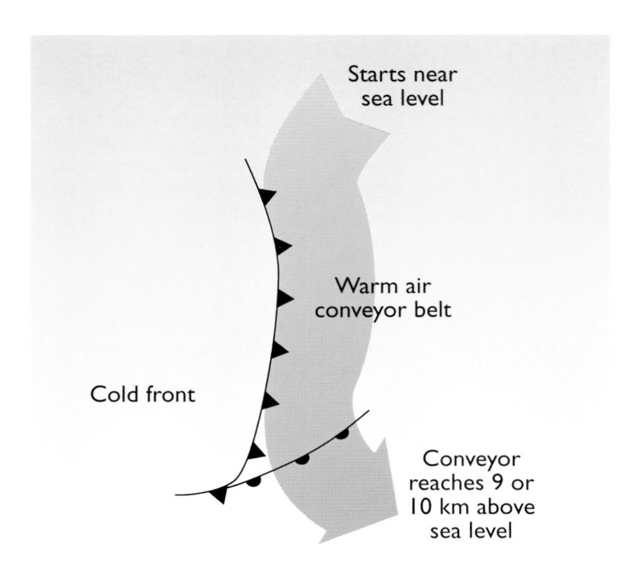

Fig. 5.2. Much of the upward motion causing cloud and rain with a cold front is caused by a warm conveyor belt parallel to the front and just ahead of it. This starts at low levels near the tropics and gradually rises to about 10 km as it moves poleward before turning away from the front and sinking.

other (fig. 5.3), but, like many other weather phenomena, they have feedback mechanisms that accelerate their growth.

One such mechanism is the heating of the air in the warm conveyor belt when water vapour condenses to form cloud droplets and in so doing releases latent heat. This increases the temperature difference across the front between the warm and cold air.

As the temperature difference increases, the difference in pressure also begins to increase, which in turn causes the winds ahead of the front to strengthen. The resulting acceleration then gives rise to a series of vertical and horizontal air currents that increase the temperature difference across the front even faster than the original wind pattern. This process can cause a front to intensify dramatically in less than 12 hours.

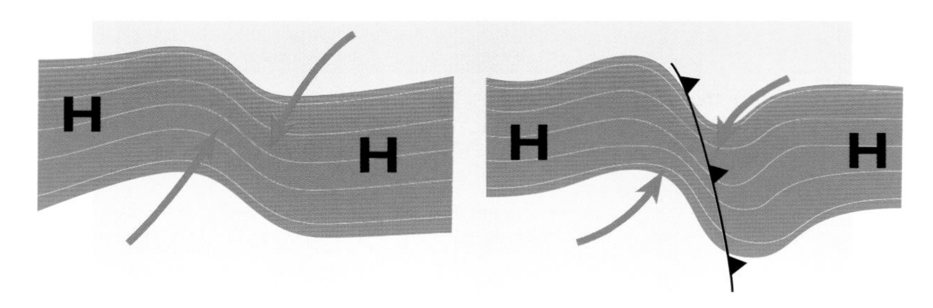

Fig. 5.3. Fronts are created when wind patterns blow warm and cold air towards each other. Various feedback processes operate to enhance the temperature difference. The lines in the diagram indicate lines of equal temperature.

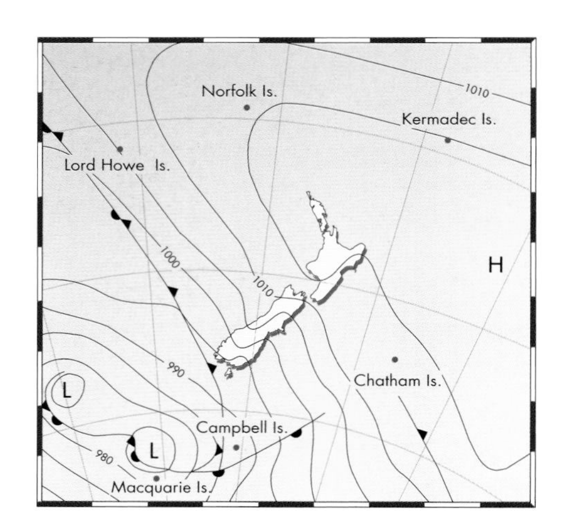

Fig. 5.4. Warm conveyor belt of air ahead of a cold front, 8 January 1994.

SOUTH ISLAND FLOOD

When the warm conveyor belt of air ahead of a cold front drives straight into the mountains of New Zealand, heavy rain and floods are often the result. A particularly intense example of this occurred in January 1994 (fig. 5.4). Torrential rain in Fiordland and Westland spilled across the Alps into western Otago and Canterbury – which might ordinarily have been safe within the rain shadow. Massive slips cut the roads to Milford and Haast, isolating hundreds of tourists who were later brought out by boat and plane. Lightning, which lasted for 16 hours along the Alps, cut electricity in many areas and started a number of fires.

One of the worst hit areas was Makarora Station above Lake Wanaka. Some 1300 sheep were drowned in a riverside paddock when the Makarora River burst its stopbank, and a bull worth $6000 was killed by lightning. Hundreds of cubic metres of silt and gravel washed down from the mountains, destroying 20 km of fencing and covering paddocks and yards.

One paradox of this situation was that rivers flowing from the mountains down to the east coast flooded near the coast, even though the coastal weather was mostly fine. The Rakaia River reached its highest recorded level, cutting State Highway 1 (fig. 5.5).

On the plus side, the hydro lakes received plenty of water and floods in the Arrow River washed fresh gravel into the river bed, producing a minor gold rush with $3000 worth of gold being taken in a few weeks.

Fig. 5.5. Rain in the Southern Alps causes floods on the Canterbury Plains.

NORTH ISLAND FLOOD

The nearer the tropics the conveyor belt starts, the warmer and moister its air and the heavier the rain and the worse the flooding. In early March 1990, a front approached the west of the North Island with a warm conveyor belt that originated inside the tropics (fig. 5.6 map). The rain ahead of the front caused floods in Taranaki and Wanganui on Saturday 10 March, and in Manawatu, Wellington and Wairarapa three days later.

In Taranaki, the Waitara and Oakura Rivers broke their banks and many people had to be evacuated. Slips closed many roads and a freight train was derailed. The Wanganui River also overflowed its banks, flooding parts of the city and rising to within one metre of the record flood of 1904.

The surface air behind the front was only marginally cooler than the air ahead of it – which had been cooled by the sea as it moved south – so there was little temperature difference across the front and it was not entirely appropriate to call it a cold front. The major difference between the air either side of the front lay in its moisture – the preceding air contained about four times as much water vapour as the air behind it.

In situations like this one, the front is sometimes designated as an occlusion, and

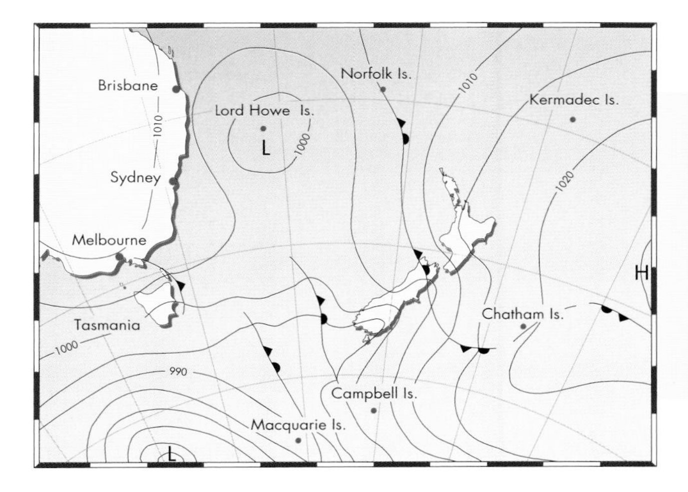

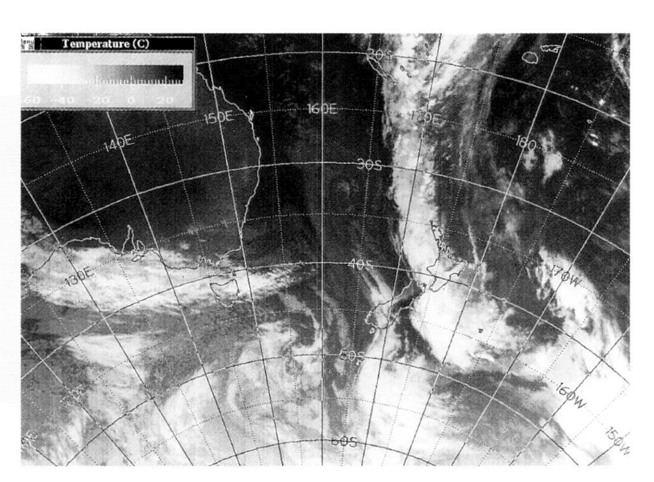

Fig. 5.6. 10 March 1990

marked on weather maps with both triangles and half moons. Traditionally, an occlusion was said to have occurred when a cold front caught up with a warm front ahead of it, and so doing forced the warm air to rise. In fact, this is a rare occurrence, and the occlusion symbol is now widely used for fronts formed by other means that have no pronounced temperature contrast between the two bodies of air at the Earth's surface.

A movie loop of hourly satellite photos of this front showed rapid movement of the cloud from the tropics towards New Zealand. However, the front was referred to as slow moving because it made slow progress from west to east. This can be deduced from the isobars which are nearly parallel to it indicating that the wind is not pushing it eastwards.

Many floods involve fronts or lows becoming slow moving for twenty-four hours or so. This enables the heavy rain to last long enough in one place to cause a flood.

CONVEYOR BELT CROSSING COLD AIR

Sometimes the conveyor belt of warm, moist air crosses back over the cold front and rises over it. When this happens a narrow belt of very heavy rain and sometimes thunderstorms is created. An example of this occurred on 3 March 1996 as a cold front moved east across the North Island. Rainfalls as heavy as 30 mm an hour occurred in Auckland, Rotorua and Tauranga, causing surface flooding with water entering a number of houses.

Radar enables us to study the workings of fronts by looking at the distribution of water droplets inside the clouds. The images show that although there are similarities, no two fronts are the same. Sometimes lined patterns of strong echoes parallel to the front have been seen on the radar screen. These lines are thought to be formed by gravity waves, which are similar to ripples on a pond after someone has thrown a stone. As one line of thunderstorms pushes vigorously up into the sky, the air immediately ahead of it tends to sink, suppressing any shower cloud. This sinking air then helps the air ahead of it to rise, causing another line of showers with strong radar echoes.

SOUTHERLY BUSTER

The traditional description of fronts included a sudden large temperature and wind change as the front passed over. In fact, a gradual temperature change with a series of wind shifts over a couple of hours is more common. In the east of New Zealand, however, large temperature changes over a matter of minutes and squally wind changes can often accompany the passage of a cold front because of the influence of the mountains.

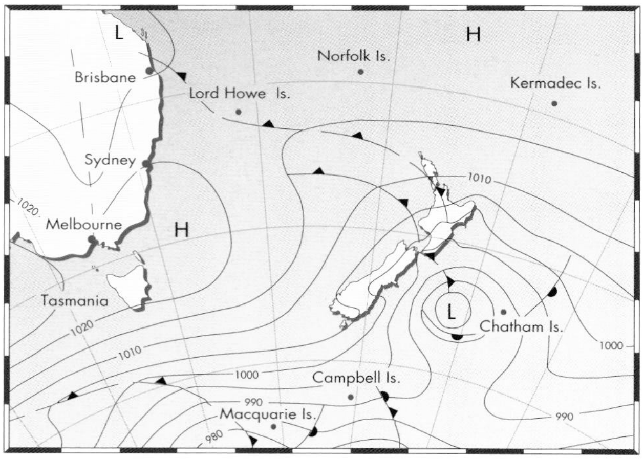

Fig. 5.7.
18 October 1994

On 18 October 1994 a cold front crossed the South Island (fig. 5.7). The warmer northwest wind ahead of it blew over the mountains and was further warmed by compression as it descended to sea level on the other side. Meanwhile, because the cold air came directly from the south, it had no mountains to rise over and so its cold temperature remained unaltered by compression.

In some places, the fall in temperature as the warm, northwest wind changed to the south with the passage of the cold front, was as much as 10°C. This means the cold air was much denser than the warm air, creating a sharp but brief rise in pressure which helped to increase the strength of the wind change. Such wind changes can be particularly pronounced where the mountains are close to the sea, such as along the Kaikoura coast. The southerly change at Kaikoura was 115 km/h (62 knots) gusting 152 km/h (82 knots) and the temperature dropped from 21°C to 11°C.

In Canterbury, these sorts of southerly changes are sometimes referred to as 'Southerly Busters', which comes from the Australian term 'Southerly Burster' used to describe a similar change that occurs along the coast of New South Wales.

Further north in Cook Strait, normally famous for its strong winds, the southerly only reached 67 km/h (36 knots) and the temperature only dropped from 15°C to 12°C. This was because the northwesterlies ahead of the front at Cook Strait had not crossed the mountains and been warmed by compression, and so the temperature and density contrast between the air either side of the front was not as great.

RAIN?

When a front crosses New Zealand, rain doesn't usually fall everywhere because of the sheltering effect of the mountains. As we saw in the last chapter (fig. 4.10), fronts crossing the South Island in a westerly flow often do not bring rain into eastern areas such as Christchurch.

If a front crossing the South Island has westerlies ahead of it and southerlies behind it, rain will fall ahead of the front in western areas, which will then clear to fine weather after the front has passed. Conversely, near the east coast, it will be fine ahead of the front, perhaps with some high cloud, and the rain will come after the southerly change (fig. 5.8).

Inland areas of the South Island, such as the Mackenzie Basin and Central Otago, can miss out on the rain from both directions because they have ranges of hills sheltering them from the southerlies as well. However, small amounts of rain can get over the ranges, and with very active fronts, such as the one in January 1994, there can be significant falls.

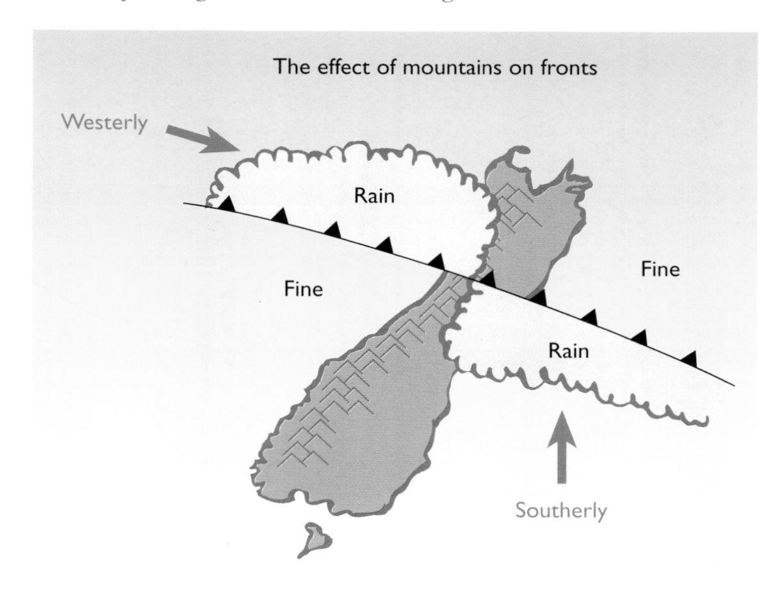

Fig. 5.8. When the wind behind a front is southerly the effect of the mountains is to reverse the sequence of weather on either side of the country. Rain falls ahead of the front in Westland, but behind the front in Canterbury.

Over the North Island, the pattern of events is similar, but because the mountains are only about half as high as the Southern Alps, it is even more likely that some rain will cross over to the sheltered side.

WARM FRONTS FROM THE SOUTHWEST

Warm fronts sometimes approach New Zealand from the southwest. This is something of a paradox, as this is also the direction which most of our cold air comes from. On these occasions, however, the warm air has swept around the southern side of an anticyclone after being over or near Australia. It travels down over the Southern Ocean in a northwest airstream before swinging around the high and moving over New Zealand in a southwest airstream.

During its journey, the air is cooled from below by the cold ocean, so when it reaches New Zealand it is not particularly warm, although it is warmer than the air it replaces.

An example of this occurred on 26 May 1996 (fig. 5.9). The temperature at Stewart Island rose from 4°C to 9°C over four hours ahead of the southwest wind change and the relative humidity rose from 60% to 97%. Light rain fell for five hours ahead of the front with a total of 2.2 mm. Similar temperature changes happened further north, but rainfall varied, with most inland areas having less than 1 mm. The rise in pressure behind the front, combined with the trough east of New Zealand, brought a gale force southerly change up the east coast of both islands as the air accelerated between the high and low pressures. In Wellington, the Cook Strait ferries delayed their entry into the harbour by several hours until the wind and waves had eased.

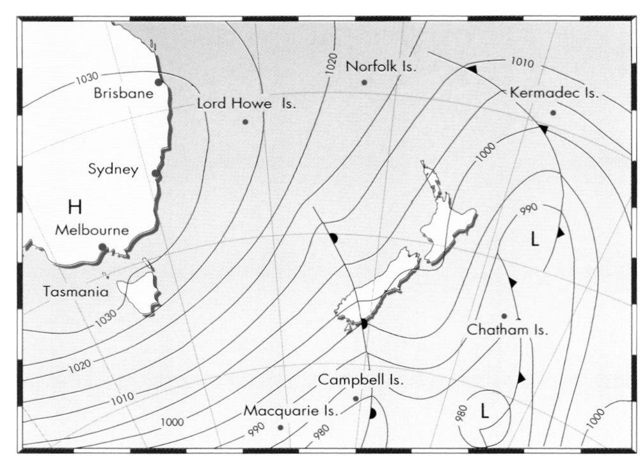

Fig. 5.9. 26 May 1996

ANOTHER PARADOX

Sometimes when a cold front passes over the temperature will actually rise. This often happens in eastern areas on winter mornings when a frost has formed during the night. The mountains provide shelter from the rain and cloud ahead of the front and, if the westerly flow is not too strong, the warm föhn wind coming down from the mountains is unable to penetrate all the way to sea level. Consequently, with clear skies and light winds during the night, the temperature drops below freezing in a shallow layer of air next to the ground and a frost forms. Then when the front rushes through, the frosty air several metres deep is replaced by air that is comparatively warmer at ground level, although colder immediately above.

CHAPTER 6

Frosts and fogs

FROST

Even though it may not always feel so, the Earth is a warm planet and constantly radiates heat. So do our bodies. You can feel this by holding your hand a couple of centimetres from your cheek. Heat radiating from the palm of your hand makes the skin on your face feel warm. All things radiate energy and would only stop if they were cooled to minus 273°C, which is the coldest possible temperature and known as absolute zero.

When a molecule at the Earth's surface radiates heat, the ray of energy may travel some distance past millions of air molecules before it hits one and is absorbed. A more direct form of heat transfer called conduction occurs when two molecules collide. Energy is then conducted from the faster moving molecule to the slower.

The Earth radiates heat both day and night, but we only notice a loss at night when the outgoing radiation isn't being countered by incoming radiation from the sun. As the ground loses heat, it cools a thin layer of air next to it by conduction; and as that thin layer cools, its relative humidity goes up. If the thin layer of air reaches 100% humidity – its saturation point – some of its water vapour will then begin to condense onto grass and other surfaces, forming drops of dew. If the temperature near the ground drops below zero before dew is able to form, then ice crystals will grow forming frost instead.

A heavy frost in North Canterbury, near Cheviot.

On nights when there is a blanket of cloud, the drop in temperature is small because the clouds re-radiate much of the lost heat back to the Earth's surface. If there are no clouds and conditions are windy, the surface air temperature may not drop much because the air close to the ground is continually being mixed with air from above. However, the ground itself becomes cold – and so do parked cars, which may be hard to start.

But in the absence of a wind strong enough for mixing to occur, a thin layer of cold air a few metres thick will develop over the land. Because the cold air is denser than the surrounding air, it will slide downhill and collect at the bottom of valleys. The temperature at the base of one of these pools of cold air can sometimes be 15°C colder than on a nearby hill top. In this way frost can affect specific areas while leaving the surroundings untouched.

Cold air sliding downhill is known as a katabatic wind and can be easily duplicated on a table top using a tray of ice cubes. Tilt the tray at an angle of about 45 degrees with its edge lined up with the table edge. Wait 30 seconds for the cold air to start sliding down the ice surface. Close your eyes and move your hand along the table edge. As your hand passes the end of the ice tray you will feel a small cold wind sliding downhill. In Antarctica, this effect creates katabatic winds as strong as 150 km/h.

Frosts are more frequent and intense in winter because the nights are longer allowing the Earth to lose more heat. They are also more intense if light winds and clear skies occur after a particularly cold southerly when the air has low humidity. If the humidity level was high, fog or cloud would form instead.

In winter, if light winds and clear skies last for some time, then the days become progressively colder and the frosts each night become more intense. This is because the weak warming from the low-angled sunlight during the brief daylight hours will not make up for the cooling during the long nights. The heat loss is accentuated if the ground is already covered in snow or ice because the surface will reflect the sunlight straight back into space – which is why people on ski-fields can get sunburnt under their chins!

FOG

Fog is simply cloud resting on the Earth's surface. It forms when humid air next to the ground is cooled enough to reach its saturation point, so that the water vapour then starts to condense into liquid droplets. The commonest condition for this is a clear night with light winds. The ground then cools as heat radiates away into space and in turn cools the air from below.

This, as we saw in the previous section, is also the recipe for dew – and indeed if the ground is cold and dry enough, the air will lose some moisture to the Earth's surface as dew. However, once the ground is wet (or if it is already wet), fog will develop after a couple of hours. If the ground temperature is below zero before the air's water vapour starts to condense, frost will form instead and the formation of fog is likely to take another six hours.

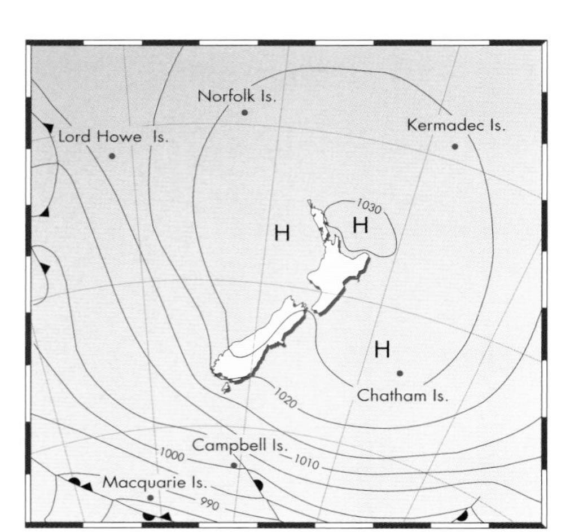

Fig. 6.1. Fog in Christchurch, 18 July 1992.

With no wind, the fog will often be only a metre or so thick and will only form over favoured areas such as grass. But if the wind is around one metre per second, then the air cooling from the ground will be gently mixed through a deeper layer of the atmosphere and the resulting fog layer can be 30 metres or more thick.

Forecasting the difference between a wind of one metre per second and no wind at all is well-nigh impossible, and even measuring it at all is difficult. Anemometers (devices that measure wind speed) are designed to withstand wind speeds of more than 200 km/h and are not sensitive to tiny fluctuations around zero.

At the other extreme, however, if the wind is too strong, then the air cooled by the ground will be further mixed with the warmer air above and either cloud will form some way above the ground or the skies will remain clear.

Another way that warm, humid air can cool enough to reach its saturation point is when it blows over a cold surface. This often happens in Christchurch when a northeast wind blows air from the relatively warm sea over the cold land during the night (fig. 6.1). A similar scenario of warm, humid air over a cold surface also happens when a cold ocean current meets a warm one. For example, fog occurs about 120 days a year east of Canada where the Labrador Current meets the Gulf Stream, which is about 17°C warmer.

Fog formed in this way sometimes turns up in Wellington in the middle of the day riding on a light southerly. This happens when warm, humid air travels towards New Zealand from near the tropics in a northeast airstream. This airstream changes to the southeast as it turns the corner near Cape Palliser and heads through Cook Strait. Sea surface temperatures near Wairarapa and Cook Strait are much cooler than further north, causing fog to develop as the warm air is cooled by contact with the colder water (fig. 6.2). The fog can last for more than 24 hours, closing Wellington Airport and causing major disruptions to aviation.

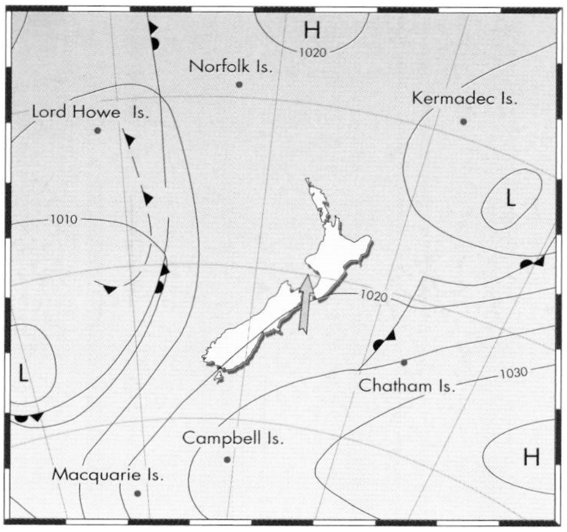

Fog can also form when cold air moves over relatively warm water. This is why fog often forms over swamps. Air cooled at night over nearby land flows downhill to the lowest ground – which is often also the location of a swamp. Swamp water then evaporates into the cold air which also gains warmth from the water surface. The warming starts to make the air less dense than the air above it, allowing it to rise and mix. And as it mixes with the air above, it cools to its saturation point and condensation begins, creating fog.

Fig. 6.2. Fog in Wellington, 25 August 1989.

Fog formed like this is common in polar waters when air sliding off ice sheets at temperatures well below zero moves over sea water a few degrees above zero. This sort of fog is known as Arctic Sea Smoke and is the same as the fog that forms over a hot bath in a cool bathroom.

Freezing Fog If the temperature of a fog is below zero, the tiny water droplets remain liquid but will freeze if they come into contact with any object like a tree or a fence (fig. 6.3). This is known as freezing fog and when it slides slowly past an object like a tree it can deposit ice many centimetres thick on the side facing into the wind.

The ice formed in this fashion is known as rime ice and is a serious danger to trawlers

Fig. 6.3. Rime ice on trees in Central Otago.

Fog fills the twisting and turning course of the Whanganui River, near Taumarunui.

operating in subzero temperatures. When spray comes into contact with the freezing metal, it instantly freezes into rime ice, quickly building up and making the vessel top heavy. The only way to remove rime ice is by bashing it off, and if this isn't done fast enough the vessel may turn turtle.

If the air temperature is below about minus 30°C, the fog will be composed of ice crystals. It is then very thin and known as diamond dust because of the way each ice crystal sparkles in the sunlight. In his book *Imperium*, Ryszard Kapuscinski tells a remarkable story about diamond dust told to him by a nine year old girl in the town of Yakutsk, Siberia. In winter, when the temperature falls well below zero, the fog that forms sticks to people's clothing so that they carve out a tunnel as they walk through it. The girl could recognise which tunnels had been made by her friends and teachers as they walked to school from their shape and direction. Some tunnels were very crooked and stopped abruptly without reaching any buildings: these had been made by drunks who had fallen down and frozen to death.

Clearing Although many fogs evaporate due to heating by the sun during the morning, there are a number of other ways they can clear. At Christchurch airport, for example, nearly half of the fogs clear before dawn. This is either because of wind changes blowing the fog away, or there being insufficient moisture to replenish the fog as it slowly falls to the ground, or warming from a cloud layer that moves over the top of the fog and evaporates it off (fig. 6.4).

Lost in the Fog The effect of fog in depriving us of our sight is so dramatic that it has entered our language as an image for confusion and uncertainty. The word 'nebulous' derives from the Latin for fog, and we use phrases such as 'a fog of uncertainty'. One of the oldest works of Western literature is Homer's *Iliad*, written almost 3000 years ago. In it the author describes a warrior's death from a sword thrust as "a mist of darkness closed over both eyes".

How fog clears

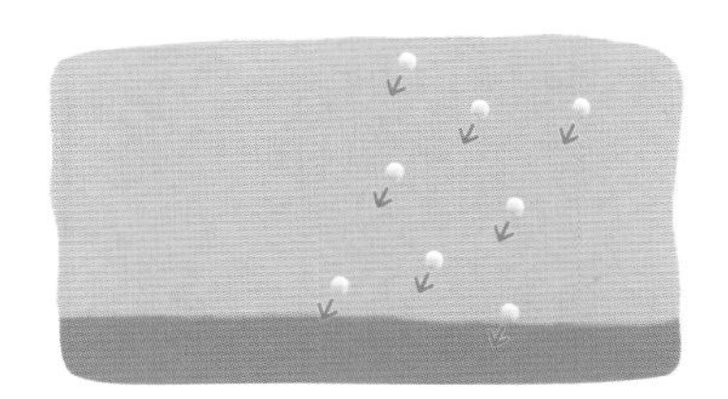

Fig. 6.4. Under the pull of gravity, fog droplets fall slowly to the ground over several hours. If fog does not keep forming at the top of the layer this will cause the fog to clear.

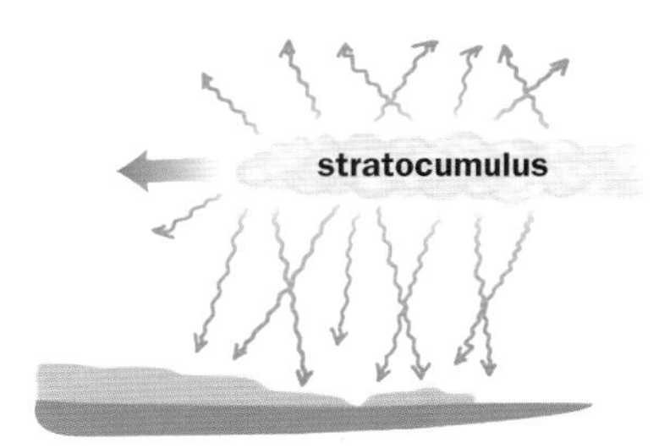

When a blanket of stratocumulus moves over the fog the infrared heat it emits can warm the fog enough for it to evaporate.

The difference between mist and fog is one of definition. In mist visibility is greater than 1000 m; in fog it is less.

Some of New Zealand's worst shipping disasters have occurred in fog. In 1894, the ship *Wairarapa*, travelling from Sydney to Auckland and sailing at full speed in thick fog, struck a steep cliff on Great Barrier Island and quickly sank. Of the 251 passengers and crew, 121 were drowned. Eight years later the *Elingamite*, also on the Sydney to Auckland run, struck West Island in the Three Kings group and sank in 20 minutes killing 45 of the 194 passengers and crew. The invention of radar and satellite navigation has now made this sort of accident far less likely.

One of the world's worst airline accidents occurred in thick fog at Teneriffe Airport in the Canary Islands in 1977 when one jumbo jet taxied into the path of another that was taking off, killing 570 people.

Fog is also responsible for some appalling accidents on motorways. Cars travelling at high speed on a fine sunny morning can turn a corner and suddenly be in thick fog. If they manage to slow down or stop without hitting anything, they then become sitting ducks for the traffic bearing down on them. This is particularly a problem in the Waikato, but is far worse on busy roads in Europe where pile-ups of as many as 120 vehicles have occurred.

The inter-island ferry *Wahine* hit Wellington's Pipitea Wharf after becoming lost in fog in 1936.

Heavy fog in Auckland on the morning of 24 May 1942, served a double purpose: it hid ships in the harbour from a Japanese reconnaissance plane operating from a submarine waiting offshore, and it hid the plane from ground observers. Susumo Ito, the pilot of the float-plane, records that he became lost in the fog. However, Auckland airport's lights were turned on, possibly in response to sounds of his lost plane circling, and Ito was able to regain his bearings and retrace his path to the submarine.

Fog forms more easily in dirty air and can combine with air pollutants – sometimes with lethal consequences. In December 1952, London had a fog that lasted almost five days and was estimated to have contributed to 4000 deaths, mostly from bronchitis and pneumonia. Breathing in this particular fog actually caused pain as a portion of the 1000 tons of dirt particles suspended in the London air was sulphur dioxide, which combines with water droplets and oxygen to form sulphuric acid!

However, fog can also be a lifegiver. In deserts situated next to cold ocean currents, although rain almost never falls, fog is common. Plants living in such conditions are able to take in water through their leaves rather than their roots, and animals lick the fog droplets that collect on their bodies. In the Namib Desert of southern Africa, there is a beetle which stands on its front legs with its back to the wind so that the fog droplets that trickle down from its back are channelled into its mouth.

A CHILLING TALE

A spectacular frost occurred in Otago in July 1991 when overnight air temperatures dropped below minus 15°C in some inland places for days in succession (fig. 6.5). Beer froze in pubs, diesel turned to sludge, water pipes burst and rabbits with frostbitten ears were accused, on national TV, of cannibalism.

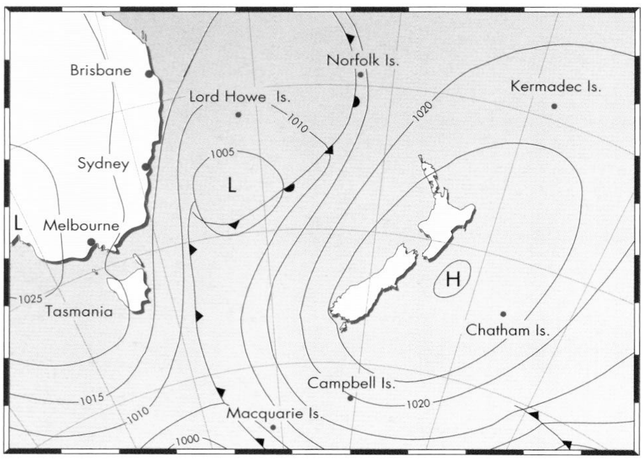

Fig. 6.5. Frost in Otago, 15 July 1991.

Icy roads caused dozens of accidents. One car skidded into Otago Harbour at Blanket Bay, with its only occupant being rescued by two teenagers who braved the cold and swam out to the vehicle.

At Mount Ida station, poledale sheep became stuck to the ground during the night when their urine froze. In the morning they had to be pulled free, leaving tufts of wool sticking to the ground.

Metal surfaces such as gates became so cold that skin would freeze on contact and be torn off if the hand was pulled away. In many places, frozen pipes meant that drinking water had to be taken from streams – once their coverings of ice had been broken.

After days of frost, freezing fog finally developed and filled the valleys for days. Everything became covered in such beautiful rime ice that there was a minor tourist boom.

All this is a good example of a blocking situation when a large, stationary high splits the prevailing westerly winds forcing them to flow either side of it. The weather map at the time featured a large, slow moving high over the South Island which stayed for a week.

Before the high arrived over the country, the air over Otago had moved up from near Antarctica in a southerly. It was initially dry with low humidity, so fog did not form on the first night the high was in place. However, the frost deepened with each successive night until the air became so cold it finally reached its saturation point (100% humidity) and freezing fog began.

Because the winter sunlight was too weak to warm things up, a warm wind was needed to bring relief. After two weeks of exceptionally cold weather, a warm northerly airstream finally put an end to the cold snap, but not before some rain had fallen. Since the ground was already so cold, the rain froze as it landed and many roads were coated in a sheet of clear ice, resulting in another spate of accidents.

Some consolation could perhaps be taken from an international survey which shows that cold weather is associated with a reduced crime rate.

CHAPTER 7

Thunderstorms, hail and tornadoes

In the early morning of 23 February 1995, the east branch of the Motueka river, normally about a third of a metre deep, rose to over five metres, washing away a hut where two Department of Conservation hunters were staying. The body of one was recovered 4 km downstream and the other was found three years later. The hut floor was found 10 km downstream and the pack of one of the men was located more than four metres up a tree.

Nelson had been subjected to heavy rain in a humid, northerly airstream ahead of a front which crossed the country late on 22 February (fig. 7.1). But the situation was greatly exacerbated by a line of thunderstorms that developed rapidly behind the front around midnight. Rainfalls of 40–55 mm were recorded in an hour at Nelson and Woodbourne airports, as well as in the Richmond Range as the thunderstorms passed over.

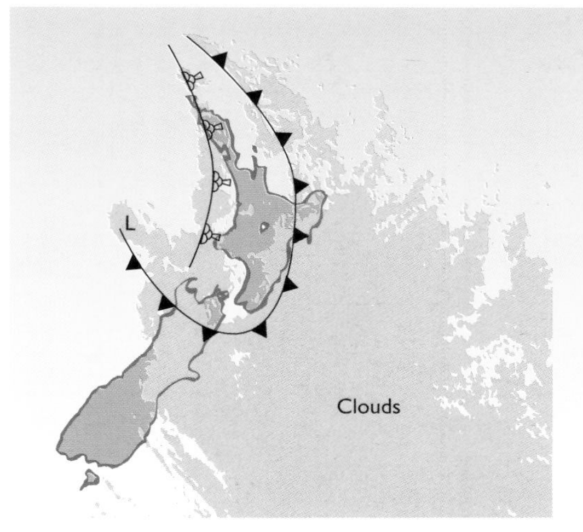

Fig. 7.1. Rapidly developing thunderstorms behind a cold front caused fatal flash floods near Nelson on 23 February 1995.

Such high intensity rainfall caused widespread flash flooding, and water entered many houses and shops in Blenheim and Nelson. Repairs to state highways affected by scouring, washouts and slips over the northern half of the South Island were estimated by Transit New Zealand to cost around half a million dollars.

In terms of human lives, however, it was a long way short of the worst flash flood in New Zealand's history. That occurred in the Kopuawhara stream between Gisborne and Wairoa in February 1938. In the middle of a Saturday night, the stream rose rapidly by five metres after thunderstorms in the coastal hills. Carrying logs and rolling boulders, it became an 80 m wide torrent, inundating 47 miners' huts in a railway construction camp. Twenty-one people drowned.

Many survivors had dramatic escapes (fig. 7.2). One man, R. Blair, woken by a friend's shouts, had barely got to his feet when his hut collapsed around him. Struggling out from under the debris, he found himself up to his neck in water but managed to grab onto a log. As it floated downstream, the log approached a 4.5 ton truck with 11 men on it. They called for Blair to join them, but he had no way of doing so; then the truck rolled over and the men were swept away.

The log continued downstream and bumped into the cookhouse, which had lost a wall but was still standing. Blair managed to clamber onto its roof and join the dozen others seeking refuge there. Another hut floated by with a man clinging to its roof and he also managed to leap across to the cookhouse. One of the men was lost when he went

Fig. 7.2. Before and after: a tragic flash flood at Kopuawhara in 1938 claimed 21 lives.

into the water to search for a waitress whose hut had been nearby. The only woman to die in the tragedy, she had probably already been swept away.

By this stage the cookhouse was starting to collapse and the men jumped across to the roof of the caterer's quarters. A number of others gradually joined them there, including the caterer and his wife, who were helped up by three men who climbed down to rescue them. All but one of the people on the roof survived.

The flow of water measured in the nearby Mangakotukutuku stream was so large it corresponded to a rainfall of 130 mm in an hour over its catchment area. In another stream, north of Gisborne, the flood rose to a depth of almost 20 m. Debris was found tangled in the top of a telephone pole.

High intensity rain is one of the lethal weapons thunderstorms have in their bag of tricks. Lightning, hail, tornadoes and downbursts are among the others. All of these are by-products of the strong upward air motion that is one of the essential features of the giant cumulonimbus clouds (fig. 7.3) that cause thunderstorms. Vertical wind speeds of 185 km/h (100 knots) have been measured in thunder-storms overseas. Although such speeds may not be reached inside New Zealand thunderstorms, updraughts of more than 93 km/h (50 knots) have been estimated by the size of some of the hailstones which have occurred.

Strong upward air motion causes very rapid production of liquid water from water vapour. This means that the release of latent heat as the water vapour condenses is extremely large. In a thunderstorm raining at the rate of 50 mm an hour, the energy released in one hour would be enough to meet the power demands of an average household for about 8000 years. By heating the air inside the cumulonimbus cloud, latent heat plays a vital role in driving the upward air motion by increasing the buoyancy of the rising air with respect to the surrounding air outside the cloud.

The word buoyant may conjure up the image of a rubber ball floating peacefully on top of the water, but in the case of a thunderstorm it is more appropriate to think in terms of a rubber ball held under water ... then released, so that it shoots violently upwards into the air.

HAIL

Hailstones grow in the parts of cumulonimbus clouds where there is an abundant supply of liquid water droplets at below zero temperatures. These liquid droplets then freeze as soon as they touch an ice particle.

The ice particles themselves will have formed on a freezing nucleus such as a speck of dust. Occasionally, insects may act as freezing nuclei and become entombed as the hailstone grows around them. Dead flies have been found inside hailstones in New Zealand. In Iowa in 1882, live frogs were found inside two hailstones, and a 15 cm gopher turtle was found inside a hailstone near Vicksburg, Mississippi in 1894. These animals would probably have been lifted into the sky by tornadoes.

As the liquid droplets freeze onto the growing hailstone, air is often trapped between

Cumulonimbus
Cumulonimbus clouds are the most violent clouds in the atmosphere, bringing thunder, lightning, tornadoes, hail and torrential rain. All of these are caused by the strong upward motions of air inside these clouds, which can reach 100 km/h or more. Cold air is denser than warm air, so a bubble of warm air surrounded by cold air is buoyant, and rises in the same way that a ball held under water will surge to the surface when released. Once clouds begin forming in rising air, the release of heat by condensing water vapour gives an enormous boost to its buoyancy. The base of these clouds is low, but they may tower up to ten kilometres.

Cold downdraught
When cold dry air mixes in through the side of a column of cumulonimbus, some cloud droplets evaporate into it. This further cools the air, making it denser and causing it to sink. The cold air accelerates downwards and hits the ground, much like water poured from a height onto the floor. These 'downbursts' are hazardous to aircraft during take off and landing.

Warm updraught
The warm updraught is the powerhouse of the cumulonimbus because it is here that enormous amounts of heat are released by the condensation of thousands of tonnes of water vapour to a liquid or solid state. In a thunderstorm producing 50 mm of rain in an hour, the heat released would be enough to meet the energy needs of a typical household for 8000 years!

Upper level airflow

Anvil

Tornado
In extreme cases, the violent updraughts in a large cumulonimbus begin to rotate, and may form a narrow funnel of rapidly spinning air known as a tornado.

Rain
Rising air is cooled by expansion, causing water vapour to condense to liquid cloud droplets, which then combine to form rain. Only about a fifth of the liquid water in a cumulonimbus cloud reaches the ground as rain. The rest evaporates either on the way down or as dry air is mixed in through the sides of the cloud.

Lightning
Lightning is the result of a build-up of static electricity caused by strong vertical air motions inside a cumulonimbus cloud.

Fig. 7.3. Cumulonimbus cloud

them. This is what gives hail its white appearance. Sometimes the hailstone will partially melt before going through another cycle of growth. Then a series of growth rings, like tree-rings, can be seen if the hailstone is sliced open.

Supercells For hail to grow to a large size, the updraught in a cumulonimbus cloud has to be strong enough to support the hailstone against the force of gravity. One place to find such strong updraughts is in a supercell cumulonimbus. These tend to develop if the winds aloft are a bit stronger than the winds near the ground, causing the cumulonimbus to grow on a tilted angle. Because of this, the vigorous updraught that creates the cloud is not destroyed by the rain it then causes. (If, on the other hand, the winds aloft are much stronger than the winds lower down, they will shear off the cloud tops and prevent the cumulus from developing into a supercell.)

A hailstorm which damaged crops in Nelson moves north over Tasman Bay in December 1996.

Falling rain will normally destroy an updraught through friction and by stealing some of the heat from the warm, rising air to partially evaporate the rain. When the tall cumulonimbus cloud is tilted, the rain falls to one side allowing the updraught to continue unhindered.

Ordinary cumulonimbus clouds have a lifetime of one hour or less, but supercells have much stronger updraughts and last longer. Hence they can hold hail within them for longer and, in fact, the hail has to be bigger and heavier to defy the powerful updraughts and fall.

Destruction Hailstorms can be immensely destructive, flattening acres of crops or stripping vegetation off trees in a few minutes. On one occasion in Hawkes Bay, particularly jagged hailstones sliced apples in two, leaving half-apples hanging from trees.

Substantial hail damage also occurs in cities. The devastating Munich hailstorm of 1984 caused an insurance loss of over US$500 million, chiefly through damage to cars.

Hail also poses a direct threat to life as a blow on the head from a large hailstone can kill. Although there are no records of people being killed in New Zealand, hailstones the size of cricket balls have been recorded here and some people have required stitches to head wounds. Animals killed by hail in New Zealand include dogs, rabbits, sheep, chickens, ducks, seagulls and, on one occasion in South Canterbury, an unlucky goldfish.

One of the worst hailstorms on record occurred in northern India in 1888. The hailstones were as large as cricket balls and 246 people were killed, along with around 1600 sheep and goats.

In New Zealand, one of the most damaging hailstorms struck Hawkes Bay on 2 March 1994 (fig. 7.4). Hail the size of golf balls destroyed most of the fruit in hundreds of orchards. The cost of the damage was over $50 million and an estimated 1000 jobs were lost.

The storm was caused by an intense trough of low pressure with record low temperatures in the middle atmosphere around 6000 m. As the cold air spilled over the western ranges of Hawkes Bay, it encountered warm, humid surface air that had been trapped up against the hills by light, northeast winds. Because the difference in temperature between these two air masses was so large, the upward air motion they triggered was particularly vigorous, resulting in a hailstorm of extreme violence.

Crop farmers and orchardists have long sought to mitigate the destructive properties of hail. Artillery shells filled with cloud-seeding chemicals are fired into thunderstorms in Russia in an effort to create lots of little hailstones before the cloud makes large damaging ones. Hail cannon that fire a blank charge have been used in many countries, including New Zealand, in the hope that the shock wave will break up the hailstones. However, this method has been compared by one expert to taking a block of ice from the fridge and trying to break it by yelling at it.

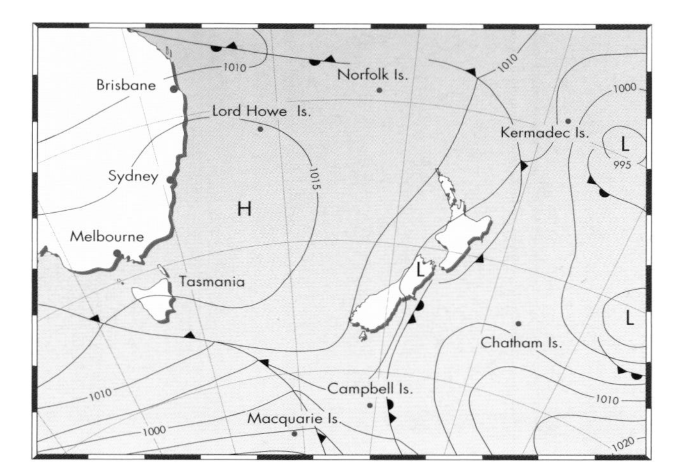

Fig. 7.4. Hailstorm in Hawkes Bay, 2 March 1994.

Although guns have been fired at the weather for hundreds of years, attempts at hail prevention go even further back. Church bells were rung as early as the eighth century to exorcise the evil spirits in clouds and many medieval bells bear the inscription *Fulgura frango* – 'I break up the lightning'. A risky claim, since bell towers are often prime targets for lightning. A medieval scholar has noted that over a 33 year period, 386 lightning strikes on church towers killed 103 bell ringers. Small wonder that arrows were sometimes shot into the clouds, perhaps to vent frustration as much as to skewer evil spirits.

Even in fine weather we are not completely safe; large chunks of ice have been known to fall from a blue sky. The most likely explanation for this is that ice formed on an aircraft only fell off after the plane had left the cloud far behind.

In New Zealand, if all sizes of hail are taken into account, the greatest frequency of hailstorms occur in western areas in winter and spring. But if only severe storms with large hailstones are considered, then the greatest frequency is in eastern districts in spring and summer. Storms that produce small hail often affect large areas at once, but the most severe storms have narrow paths of destruction and leave most places unscathed. For this reason, they can only be precisely predicted a few hours ahead, with the aid of weather radar.

LIGHTNING

Lightning is the discharge of static electricity that has built up inside a cumulonimbus cloud. It travels between different parts of the cloud, or between the cloud and the ground. If the jagged line of the lightning stroke can be clearly seen, it is called forked lightning. If the lightning stroke is inside the cloud so the cloud lights up like a frosted light bulb, it is sheet lightning.

Static electricity builds up inside clouds in a variety of ways still not completely understood. One theory involves the way in which large water droplets break up inside clouds. As they fall, these drops flatten out into a disc shape, then the centre of the disc thins and bulges up, a bit like a parachute. At the stage just before they break up, the droplets have become polarised with a negative electric charge on the end nearest the

Lightning discharges heat the air to 30,000°C (five times the temperature of the sun's surface) causing the explosive expansion of air that we hear as thunder.

Earth and a positive charge on the thin roof of the parachute.

When the drop fragments, the thin roof breaks into small droplets that are positively charged, while the lower part breaks into mostly larger droplets that carry a negative charge (fig. 7.5). Studies have indicated that a lot of the charge separation takes place near the freezing level, so it is likely that collisions involving ice particles also play a role.

Once enough droplets or ice particles have become separated, the larger cloud particles have a mostly negative charge and smaller particles a mostly positive charge. The strong updraughts inside the cumulonimbus cloud then continue the separation process by carrying the smaller positive particles thousands of metres higher than the larger negative particles, creating a powerful electric field.

There are now two areas of opposite charge within the cloud, but electricity is at first unable to flow across and neutralise them because the air in between acts as an insulator. For it to be able to conduct, some of the air molecules must become ionised by a strong electric field – such as the one we have just built up inside a cumulonimbus cloud.

If, however, the negative charge near the base of the cloud is strong enough, it can induce a strong positive charge on the Earth below. This second area of opposite charge may often be closer than the positive charge higher up in the cloud, in which case, in order for a discharge to take place, less air will need to be ionised below the cloud than within it.

Once the electric field in the air below or within the cloud has increased to about one million volts per metre, the air is ionised sufficiently to conduct an electrical discharge. In the case of the discharge trying to reach the ground, electrons begin to cascade downward in short 100 m steps, each lasting less than a thousandth of a second. Known as the step leader, these downward strokes are often invisible, but once one reaches about 100 m above the ground, a return stroke comes up from the Earth to meet it. This upward, return stroke is the more powerful and can reach 100,000 amps. It may last only one millionth of a second, but it is this flash which is usually seen.

The temperature in a lightning stroke can reach as high as 30,000°C – more than five times the temperature of the surface of the sun! Intense heating along the discharge path causes an explosive expansion of the air, which we hear as thunder.

But because the speed of sound is much slower than the speed of light, it is normal to see the lightning flash before hearing the thunder. The length of time before the thunder arrives is a measure of how far away the lightning is. A wait of three seconds between the lightning flash and the thunder means the lightning has discharged 1 km away.

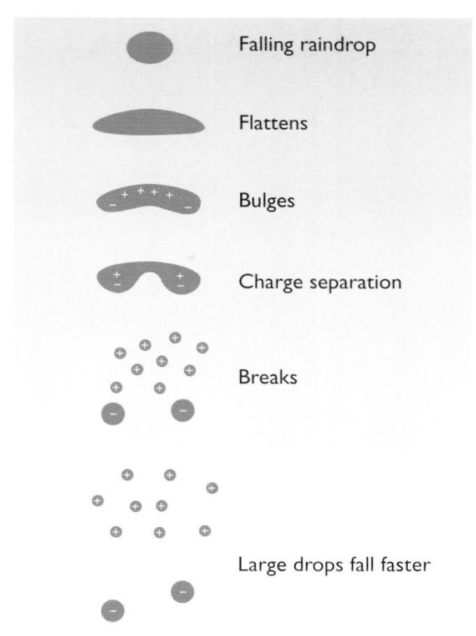

Fig. 7.5. One of the ways electrical charges become separated in thunderstorm clouds is through the break-up of large raindrops.

Lightning Strikes When lightning passes through something solid, like a tree, all the moisture along its path is instantaneously vaporised causing the object to explode (fig. 7.6). This is also how lightning sometimes ploughs up the ground. Some years ago when the search was on for kakapo in the mountains of northwest Nelson, marks made on the ridge tops by lightning were mistaken for kakapo trails.

In the United States, lightning kills about 100 people a year, frequently on golf courses; whereas in New Zealand, deaths from lightning only occur about once every five years.

On 28 April 1990, Russell Cullen had the narrowest of escapes when he was fishing with his friend Bob Watson in an aluminium dinghy off Warkworth. Late in the afternoon the sky darkened as clouds moved off the land towards them. Suddenly there was a blinding flash and a deafening noise. Instinctively, Bob Watson closed his eyes. When he opened them a split second later there was light dancing along the rim of the dinghy and his mate Russell had disappeared.

Moving quickly to the bow of the dinghy, Bob looked over and saw his friend underwater. He somehow managed to drag him back on board and set about trying to revive him as another boat towed them into shore. On the wharf, volunteer firemen worked on resuscitation for a further 25 minutes before they finally got a pulse. Miraculously, Russell survived without brain damage, partly because of his immersion in cold water.

The same thunderstorm also made its presence felt over land. Hail the size of 50 cent pieces caused thousands of dollars worth of damage to horticultural crops and a small tornado demolished a barn, then speared its timbers through a car and the side of a house.

The weather map (fig. 7.7) on this occasion showed a weak front crossing the North Island. The fact that it was weak signifies that the temperature difference between the air ahead of and behind it was small and therefore the upward air motion associated with the front itself was also small. However, the relative warmth of the air behind the front plus its high humidity were key factors in the development of the thunderstorm. The other was that the polar exit of a jet stream passed over the area causing cooling in the middle atmosphere (fig. 7.8). The combination of cooling aloft with warmth and high humidity near the surface was an ideal recipe for rapid upward air motion and

Fig. 7.6. Joe Perez holds pieces of scorched wood by a huge stump left by a lightning strike in 1986.

vigorous thunderstorm development.

Another dramatic lightning strike occurred at Hamilton Boys High School in October 1987. About 20 boys were sheltering from heavy rain under a tree when a bolt of lightning struck it. Eleven were knocked to the ground and one had to be revived with mouth-to-mouth resuscitation. He suffered serious burns to his face, neck and shoulders, and especially to his elbows and the soles of his feet.

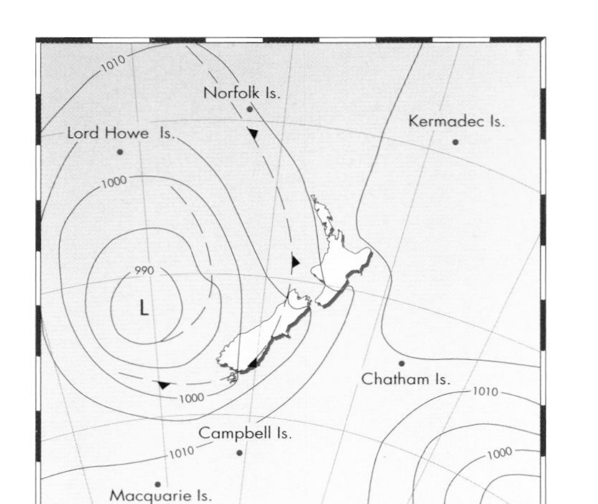

Fig. 7.7.
28 April 1990: a weak front, high humidity and warm air contribute to the development of thunderstorms.

In both of the accidents mentioned, the victims would have died but for the prompt first aid they received. In December 1986, a shepherd mustering near Taihape was less fortunate. There was no one nearby when he and his horse were struck by lightning and both were killed.

A direct hit is not the only danger from lightning. The electrical charge can travel considerable distances through fence wire, telephone lines and even plumbing. A number of people using telephones have been thrown across rooms or burnt when lightning has struck the lines.

A house in Waitara was struck by lightning in April 1992. The charge travelled through the electrical system destroying most of the light fittings and sockets, blowing the meter board clean off the wall and causing an explosion in the gas main under the house. It was only saved from fire by the fire-resistant paper under the floor.

Lightning tends to discharge through the highest point available, so mountain tops are dangerous places to be during thunderstorms. Sheltering under lone trees is also a foolish idea, as is standing in a field. Holding up short lengths of metal like fishing rods, golf clubs, or umbrellas, can also lead to a shocking experience.

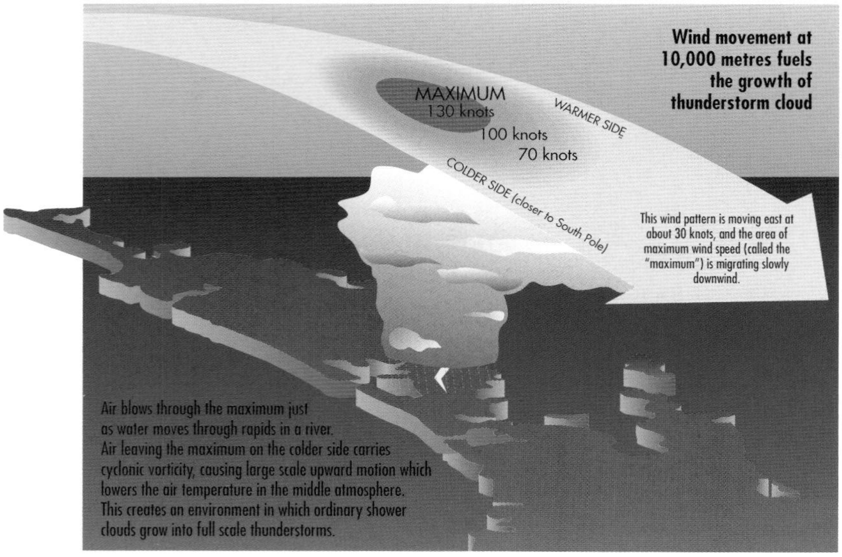

However, the chance of being hit by lightning is very small. There is a much higher chance of being nearby when someone else is hit and, as we have seen, a knowledge of first aid in such situations can be worth someone's life.

When Oscar Wilde quipped "Everyone is always talking about the weather but no one ever does anything about it" he forgot about Ben Franklin's lightning rod invention. Lightning rods are metal rods fixed to the exposed parts of buildings – often farmhouses or barns – with a wire running to the ground to conduct the lightning strikes safely to the Earth. Introduced in the 1750s, they were a great success.

Fig. 7.8. Upper air flow, 28 April 1990.

Franklin even had the good timing to invent them soon enough for James Cook to take one on the *Endeavour* on his voyage to New Zealand. Called an electrical chain, it was deployed to good effect when they were anchored in the harbour of Batavia (Jakarta) during a thunderstorm. A nearby Dutch ship had its mainmast shattered by a bolt, but when the *Endeavour* was also struck the lightning was channelled harmlessly into the sea.

Ball Lightning Among the unsolved mysteries involving lightning is the rare and exotic phenomenon known as ball lightning. Varying in size from a cricket to a soccer ball and glowing with the intensity of a domestic light bulb, ball lightning can hang suspended in the air or wander about erratically. Some observers have reported seeing "blue worms of light" writhing inside the ball. The phenomenon can end quietly or explosively, sometimes causing injury and death.

DOWNBURSTS

Mature cumulonimbus clouds pose a considerable danger to aircraft through their severe turbulence and ability to cause icing. The supercooled water they contain will freeze on contact with the leading edges of an aircraft's wings and other surfaces. Aircraft have a variety of de-icing devices, but generally try to avoid flying through cumulonimbus clouds – which is why commercial aircraft carry weather radar.

The severe turbulence within cumulonimbus clouds is caused by their strong updraughts, but is exacerbated by the fact that immediately next to an updraught there is also often a strong downdraught.

As a cumulonimbus punches up through the sky, the air pushes vigorously out from the sides of the growing tower, forming the characteristic cauliflower edge that is so distinctive in cumuliform cloud. Part of this cauliflower shape involves air from outside being rolled into the cloud. This is called entrainment.

Once this air is inside the cloud an interesting process is triggered. Because the air from outside the cloud is below 100% relative humidity, some of the cloud's rain or liquid droplets begin to evaporate into it. But evaporating the liquid droplets uses up some of the entrained air's heat and so its temperature falls. This makes the entrained air denser than the rest of the air in the cloud, and it begins to descend.

As it sinks, the air is compressed and its temperature begins to rise again. However, a rise in temperature lowers the relative humidity and therefore more liquid droplets are able to evaporate into it and the air cools further. Hence we have a runaway feedback process accelerating the cold air downwards and so a downdraught is born.

The strength of the downdraught depends on how much colder and drier the air outside the cloud is in comparison to how much water the cloud already contains.

As the cold air falls out of the cumulonimbus it acts something like water being poured from a bucket onto a floor. When it hits the surface it spreads out rapidly in all directions, forming a wind known as a gust front that can travel horizontally a long way from the parent cloud and temporarily reverse the prevailing wind.

A gust front strong enough to rip branches off trees and destroy small buildings was reported immediately before the hailstorm over Hastings on 2 March 1994 (fig. 7.4). As the cold air crashed down out of the storm, the temperature dropped from 21°C to 14°C.

When a downdraught spreads out from the base of a cumulonimbus cloud it can be extremely dangerous to aircraft taking off or landing. A headwind can suddenly become a tailwind, reducing airflow over the wings (and therefore lift) and causing an aircraft to stall or even crash.

Squalls caused by downbursts are also very dangerous to yachts.

TORNADOES

Tornadoes are violently rotating funnels of air that extend from the base of a small fraction of cumulonimbus clouds. Usually they last only a few minutes, but in rare cases they

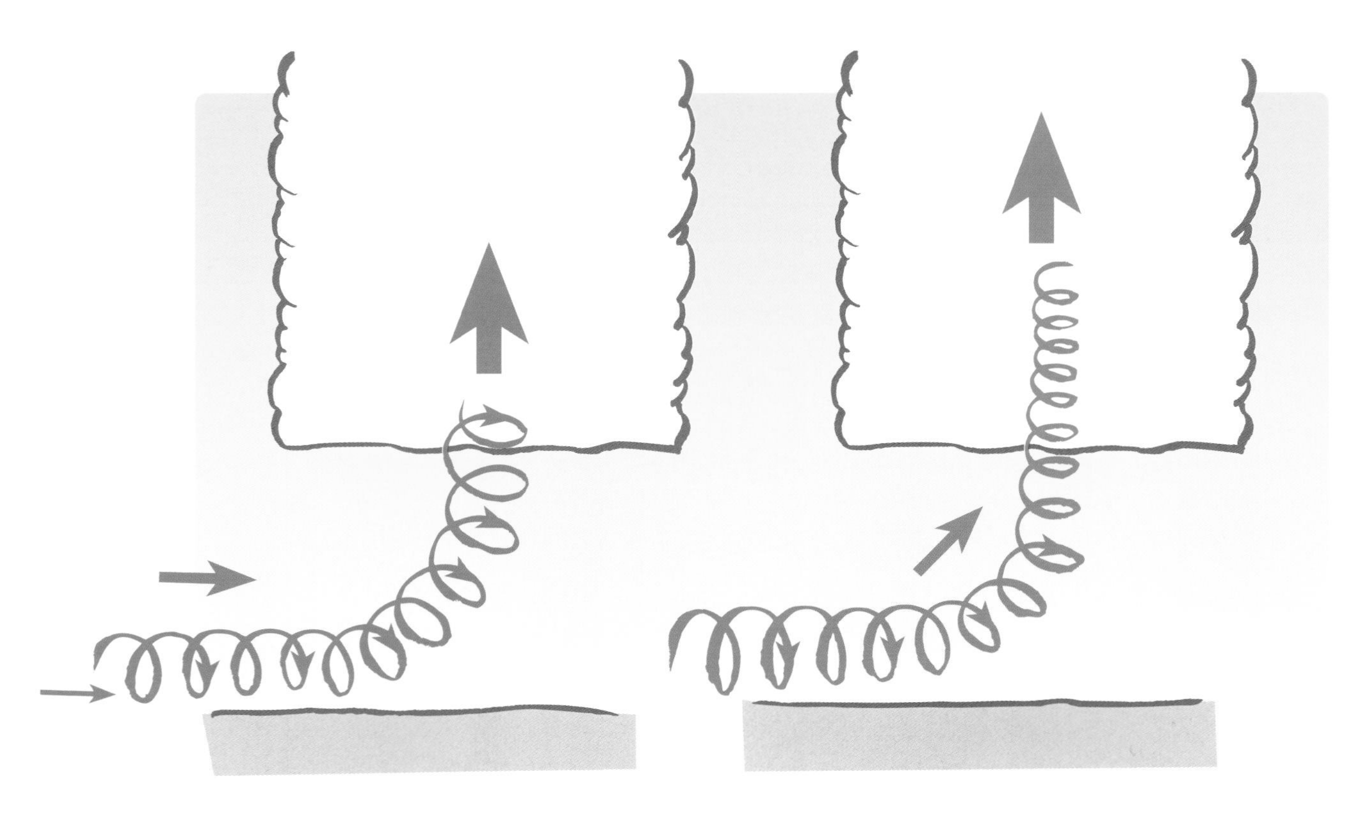

Fig. 7.9. Tornado formation: when friction slows the air next to the ground, the stronger wind above causes the air to roll over in long horizontal tubes. When lifted into a cumulonimbus cloud, the part of the tube entering the cloud first is pulled rapidly up, causing the tube to stretch and therefore narrow and spin faster. This plays an important role in tornado formation.

can last for hours. With wind speeds in excess of 200 km/h, they can be extremely destructive and are the subjects of intense study. Although their formation is not completely understood, a number of things are clear.

The initial rotation of the air often comes about because the surface wind is usually lighter than the wind above because of friction at the Earth's surface. This has the effect of causing the air to rotate about a horizontal axis – the same as a rope will do if you roll it along a table top with the flat of your hand. If this rotating cylinder of air is then lifted into a cloud, its rotation becomes orientated about a vertical axis. Again, you can see how this works by lifting one end of the rolling rope off the table top.

In order for the air rotating round a horizontal axis to tilt up vertically, a strong updraught, like those found in cumulonimbus, has to exist. When a column of air spinning about a vertical axis enters the base of a cumulonimbus, the top of the column is pulled up into the cloud faster than the bottom. This stretches the column, causing it to become narrower or contract about its axis of rotation. It's a bit like stretching a rubber band – as it gets longer it also gets thinner (fig. 7.9).

But the main point with stretching the spinning column is that as it contracts, it spins faster – just like spinning ice-skaters when they bring their arms in close to their body. This is something you can experience for yourself if you have a swivel chair. Just extend your arms and legs out as nearly horizontal as you can manage and lean back. Now start spinning gently, then sit upright pulling your arms and legs into your body and you should feel a quick increase in your spinning.

Rotation within a supercell cumulonimbus cloud actually intensifies the updraught by protecting it from entrainment of the surrounding air. And in another of the weather's symbiotic feedback mechanisms, the updraught in turn intensifies the rotation by further stretching the spinning column. Eventually, by processes still not fully understood, this can lead to a tornado growing down from the cloud. Tornadoes usually form at the boundary between the warm air being sucked into the updraught and the cold air from a downburst that has fallen out of the cloud, so it is likely that some interaction between the warm and the cold air also contributes to their rotation.

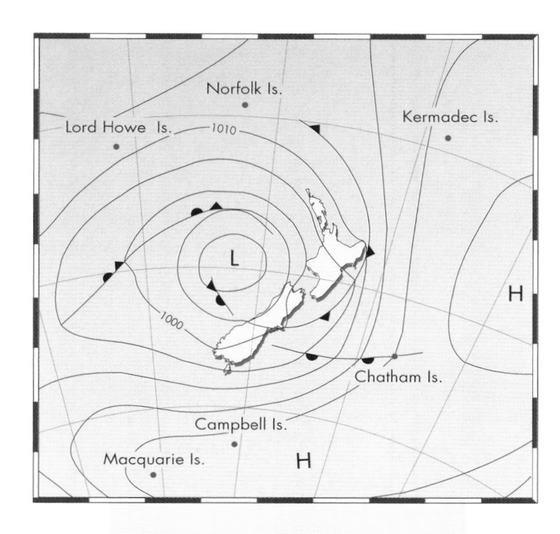

Twister! A particularly destructive tornado occurred in Taranaki on 12 August 1990. The inhabitants of Inglewood were woken at 3.30 a.m. by the sound of wind tearing buildings apart (fig. 7.10). People were thrown across rooms and hundreds of windows were blown in, yet, miraculously, no one was killed.

The luckiest survivor may have been Paul Dodds. A 400 kg steel beam was carried 100 m and lifted over a two-storey building before it crashed into the room where he was sleeping, bringing the roof down on top of him. Fortunately, several cars in the adjoining garage supported the debris and saved him from being crushed.

The morning unveiled 43 damaged buildings, trees that had been uprooted or snapped off, bodies of sheep killed by flying roofing iron and three goats that had been carried high up into the trees of a shelter belt.

One of the most remarkable tornado escape stories in New Zealand is that of Manukau dairy farmer Laurie Coe, who was sucked up into a tornado that uprooted trees and tore roofs from houses on the Awhitu peninsula south of Auckland in September 1990. Carried 100 m before being slammed down on his back in a paddock, he suffered only a stiff neck, sore back and a few minor bruises. Mr Coe had just finished milking, but his cows refused to leave the yard and make their way back to the paddock. His dog also sensed something and cowered beside him. Mr Coe drove off on his four-wheel drive farm bike to see what was worrying them. Unable to see the approaching tornado because of heavy rain, he and the bike were picked up and slammed into a fence before he was suddenly lifted into the sky. Mr Coe described it as like spinning around in a clothes dryer and said it was the most terrifying experience of his life.

The worst tornado in New Zealand's history struck Frankton and Hamilton in August 1948. Almost 150 houses were wrecked, three people killed and dozens of others injured.

Fig. 7.10. James Cunniffe in the remains of his grandparents' house in Inglewood, 12 August 1990.

But New Zealand is fortunate compared to the United States where tornadoes are more frequent and often awesome in their power. The worst tornado outbreak killed around 700 people on 18 March 1925 as it crossed the States of Missouri, Illinois and Indiana during a three and a half hour rampage.

More recently, a major tornado outbreak on 3–4 April 1974 saw 148 tornadoes across 13 states kill more than 300 people, injure over 6000, and cause damage totalling US$600 million.

Waterspouts Tornadoes occasionally form over water and are then known as waterspouts because of the column of water sucked up into them. On Cook's second voyage to New Zealand in 1773, he encountered a number of waterspouts as he sailed past Stephens Island, one passing within 50 m of the *Resolution*'s stern. Cook had heard that firing a gun into a waterspout would dissipate it. Eager to try this, he ordered that a small cannon be loaded and aimed, but records how "his people" were "very dilatory about it, and the danger was past before we could try this experiment". A debate among the officers as to which way the waterspouts were rotating was resolved when a seagull was caught in one and whirled around as it was taken up.

Strange Phenomena As well as destroying buildings and picking up cars, tornadoes are responsible for some strange phenomena. One of these is driving straw into trees. Often thought to be testimony to the brute force of the wind, there is an eyewitness account of how this really happens.

In 1953, a Michigan policeman heard a roar like a freight train and went outside to investigate. Picked up by the wind of a tornado, he found himself slammed against the wall of his house and pinned there while a tornado raged around him. To his amazement, he saw a tree being twisted round like the green top off a carrot. The tree did not snap off, but the trunk opened just like the strands of a rope will part as it is untwisted. After the tornado had passed, the policeman went up to the tree and found a stalk of grass had entered the trunk while it was twisted open and been trapped there when it snapped shut again.

On another occasion in Canada, a cow picked up by a tornado was thrown back to earth so that its horns stuck into the ground. Dug up and set back on its feet, it appeared none the worse for its experience.

Chicken plucking is another amusing trick of tornadoes, which is probably caused by the very low air pressures that exist inside them. Because the air pressure inside the hollow shaft of a feather cannot reduce as quickly as the outside air pressure, it pops out.

Perhaps the prank which tornadoes are most notorious for is their redistributing of things around the world. If a tornado goes over a pond or an estuary it can suck up a great many fish, frogs, or other animals, along with the water they were living in. Once inside the cloud, things get sorted out according to size and weight. The water goes one way and the fish another, so that the cumulonimbus cloud is able to drop all the fish of one size in one person's backyard and dump the water somewhere else.

Among the more bizarre examples of this tornado delivery service recorded around the world are: a rain of freshwater mussels in Germany in 1892; a rain of spiders in Hungary in 1922; a rain of crayfish in Florida in 1954; a rain of maggots in Acapulco in 1968; and a rain of hazelnuts in Ireland in 1867. So in this way, a cumulonimbus cloud with a tornado could literally make it rain cats and dogs!

CHAPTER 8

Snow

Snow fell to low levels over Canterbury and Otago on 8 July 1992. The heaviest falls were in the foothills where snow accumulated to depths of one metre or more over several days. Roads were closed, power lines brought down and tens of thousands of cattle and sheep were trapped.

The snow stayed on the ground for more than a week on hill country farms and stock rescue became a major operation involving hundreds of volunteers, including prison inmates, off-duty police and Lincoln University students.

Heavy snow to low levels results from the meeting of cold air from the south with warm air from the north. Cold air is unable to contain much water vapour, so if the air through the entire depth of the atmosphere has come from near Antarctica, there may be some snow to low levels, but not a great deal (fig. 8.1).

However, if a depression brings warm humid air down from the north and this is undercut by very cold air from the south, then a heavy snowfall is possible (fig. 8.2). The warm air rises over the cold and both are forced to rise by the hills. The rising air then cools by expansion and condensation begins. If the temperature of the rising air falls below about minus 10°C, ice crystals will start growing much more rapidly than liquid droplets.

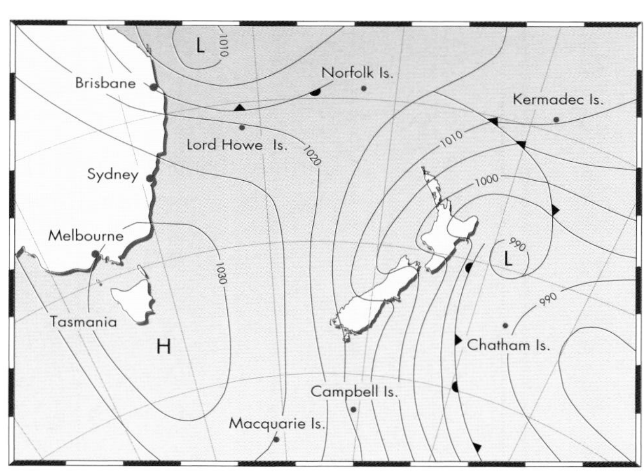

Fig. 8.1. Very cold southerly airstream 10 km deep brings air from near Antarctica over New Zealand. Heavy snow fell in the mountains, but although snow fell to very low levels in the east of both islands, the amounts were small.

The genesis of a snowflake is the same as that of a hailstone – a microscopic particle such as clay or bacteria. In air below 0°C, molecules of water latch onto these freezing nuclei and the ice crystal begins to grow. In a hailstorm, with its strong upward air motion, the hailstones grow rapidly through violent collisions with liquid droplets. But in a snowstorm the upward air motions are much gentler and the ice crystal grows as molecules of water vapour are deposited onto its delicate branches. Snowflakes are then the product of many ice crystals that have stuck together.

Many snowstorms start with a couple of hours of rain. At first the falling snowflakes melt in the air before they reach the ground. The act of melting, however, takes heat away from the air, and once its temperature has cooled to near 0°C, the snow is able to penetrate all the way to the ground. The more snow there is falling from higher levels, the faster this cooling takes place. The amount of snow reaching the ground will also increase if it falls through low level clouds where the snowflakes can scavenge some of the cloud droplets and grow bigger.

Paradoxically, once it is on the ground, the snow acts as an insulator. Trapped animals are often sheltered from the wind and their body heat can help form small snow caves. In

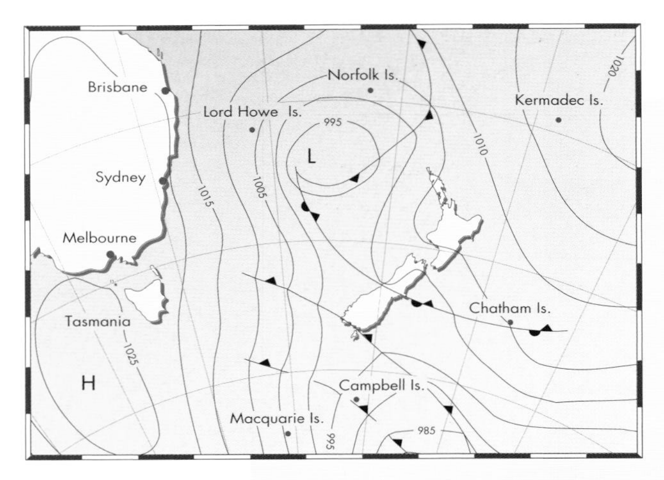

26 August 1992

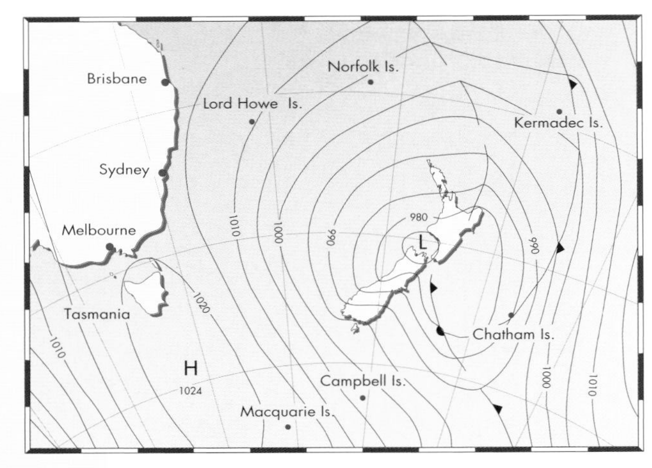

27 August 1992

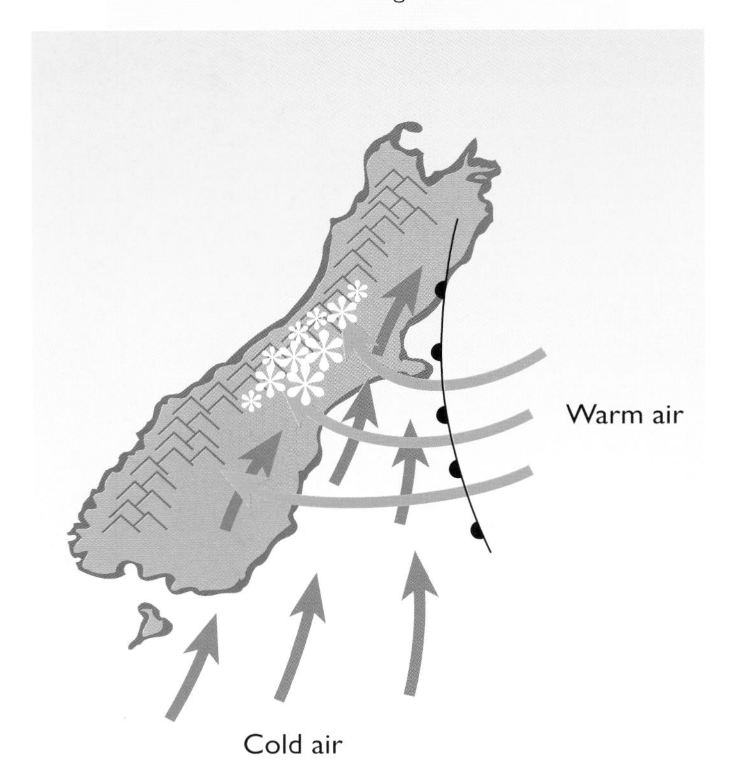

Fig. 8.2. Heavy snow occurred on 8 July and 27 August 1992 when warm humid air from the north and very cold air from the south were brought together over Canterbury by a deepening depression.

the clear nights that followed the Canterbury and Otago snowfall, temperatures dropped to minus 16°C at the snow surface, but the ground temperature stayed near 0°C.

Thaw In midwinter, the sunlight in Canterbury is not strong enough to melt snow, especially when most of the rays are being reflected away by the white surface. Warm winds or rain are necessary before thawing will take place. These finally arrived on 19 July when a warm nor'wester developed. However, the strong winds further damaged the Canterbury forests which had already lost many trees and branches to the heavy snow.

There was a fear that the thaw would lead to serious flooding, but this never eventuated because the Canterbury nor'wester is such a dry wind that much of the snow and water evaporated directly into the air. But disastrous floods have followed a number of heavy snowfalls in the past. In 1868, the Opihi River in South Canterbury ran 11 km wide when swollen with meltwater.

Nevertheless, the 1992 July snowfall had a devastating effect on Canterbury farmers (fig. 8.3). Thousands of animals died and some farmers faced helicopter bills as high as $30,000. Many of the surviving stock were in poor condition from dehydration and lack of food, having had little more than bark or wool to eat. They were prey to illnesses such as sleepy sickness, caused by lack of sugar, and staggers, caused by lack of magnesium.

More warm nor'westers were needed to promote spring growth and alleviate the need for hay, as well as to dry low-lying paddocks that had been turned to mud when stock were concentrated on them. The mud attracted large numbers of seagulls and it was feared that they would peck out the eyes of any weak sheep that collapsed.

Storms of this severity exact a far greater death toll if they occur in the middle of lambing. Tragically, in the last week of August, another severe snowstorm hit Canterbury and over a million lambs perished. This storm occurred in the same manner as the July storm, but the snow was heavier in some areas, particularly parts of Banks Peninsula where drifts of up to six metres were reported.

SNOWDRIFTS

When strong winds blow snow around, it banks up into snowdrifts, the deepest of which tend to form in hollows or sheltered places. During an exceptional snowstorm in the northeast of England in February 1941 – not much reported at the time because of wartime censorship – six trains were buried by snow in a cutting near Newcastle. About 1000 people were trapped, and they were only discovered when someone walking over the top heard their voices coming up through the snow.

WHO GETS THE SNOW?

The weather a particular place gets is partly determined by latitude and partly by the proximity of large landmasses, mountain ranges and warm or cold ocean currents. For example, Takada, on the west coast of Japan, is on a similar latitude to Auckland and only 20 m above sea level, yet it has an annual seasonal snowfall of seven metres. In the month

Fig. 8.3. The scene near Fairlie, after Canterbury's 'big snow' in July 1992, which had a devastating effect on farmers.

of January 1945, it had just under 10 m and snow covered the ground for more than four months.

Takada's snow is caused by northwest winds bringing extremely cold air from Siberia – where winter temperatures regularly fall below minus 20°C – across the relatively warm Sea of Japan. As it passes over the sea, the air picks up abundant moisture through evaporation and is heated and destabilised by the sea's warmth. Therefore, by the time it rises over the land it is primed to drop large amounts of snow.

The key thing in this situation is the large temperature difference between the cold air and the warm sea. Although cold air cannot contain a lot of water vapour, in this case the temperature difference creates such vigorous upward air motion that the cold air moves through the cloud system very rapidly, enabling significant quantities of water vapour to become snow.

The same process occurs in the United States when Arctic winds blow over the Great Lakes. Towns immediately downwind of the lakes can receive metres of snow while adjacent areas get almost none.

AVALANCHES

A snowflake falling through the air is one of the most peaceful sights in nature, but a million billion snowflakes falling together are potentially one of the most destructive. A 500,000 tonne avalanche peeling vertically off a mountainside can reach speeds of over 300 km/h. So much air is displaced by avalanches that they are often preceded by an air blast that can rip a building apart or puncture the eardrums of a bystander.

Avalanches are a danger to skiers and climbers and, in New Zealand, few years pass without one or two fatalities. But the risk to life increases if there happens to be a road through an area of high avalanche risk, such as the road through the Homer Tunnel (fig. 8.4) to Milford Sound.

Construction of the tunnel began in 1935, and there were three fatalities among the construction crew from two different avalanches in the first two years. Because of this, work was suspended during the winter, and even after the tunnel was completed the road was closed during the dangerous months of the year to avoid further accidents.

Since 1977, however, because of the increasing number of tourists wanting to visit Milford Sound, the road has been open throughout the winter and any avalanches that fell were cleared – a hazardous job, given that the conditions which cause one avalanche can often cause another soon after. In fact, the most recent fatality occurred in 1983 when two men, working to clear debris from one avalanche, saw a second coming over the top of the cliff. One man, inside a bulldozer, bent down and jammed his head under the steering clutches as he feared the cab would be sheared off, while the other crouched beside the machine.

The air blast upended the bulldozer, then seconds later the mass of the avalanche struck, driving the machine 150 m down the road and burying it. The man inside, although suspended upside down, managed to switch off the bulldozer, climb out the burst windows, then dig his way to the surface. He found the other man dead nearby.

This accident led to a change of tactics in avalanche control. Now there is an active programme to release heavy build-ups of snow with explosives while the road is closed. MetService forecasters are in contact with Works Civil Construction several times a day during periods of high risk. They forecast amounts of rain or snowfall, temperature and, most importantly, wind speed and direction at ridge top level. Because wind concentrates snow in certain places, it is sometimes called the architect of avalanches.

Fig. 8.4. An avalanche triggered by explosives roars across the entrance to the Homer Tunnel in the upper Hollyford valley.

Many avalanche situations result from new snow falling onto old and failing to bind strongly to the old surface, which may have an icy crust either from melting and refreezing or from the effects of wind or frost. Sometimes a weakness can even develop within the snowpack if the snow's surface temperature is much lower than that of the base. The temperature difference then allows some of the water molecules from ice crystals near the base to change to water vapour and move up through the snowpack and redeposit on ice crystals near the top, weakening the structure.

Avalanches release because of increases in weight on a weak layer when more snow falls or arrives with the wind, or because of temperature changes destabilising the snowpack, or even because of sound waves. Artificial release above the Milford Road is achieved by dropping explosives from helicopters.

In the European Alps, it is estimated that over 100,000 avalanches occur each year. Descriptions of avalanches go back to the third century BC when they killed some of Hannibal's elephants and soldiers as his army crossed the Alps to attack Rome. During the First World War, the front line between Italy and Austria ran through the Alps and an estimated 60,000 troops were killed by avalanches, many of them triggered deliberately.

The worst avalanche accident in mountaineering history occurred in the Soviet Pamirs in the summer of 1990 when a massive fall of ice and snow killed 43 climbers camped high on Peak Lenin. There were only two survivors.

But not all avalanches have been disastrous and there is a least one on record that was positively beneficent. Some time in the nineteenth century "a giant snowball" landed beside a poor man's cottage in Scotland. He and his family were uninjured and when they dug into it they found: "three brace of ptarmigan, six hares, four brace of grouse, a blackcock, a pheasant, three geese and two fat stags".

Fig. 8.5. The Tasman Glacier moves along at a sedate 65 cm a day, though the warming climate since the end of the Little Ice Age means the general trend for this and all other glaciers is one of retreat.

GLACIERS

Fortunately, there is a more leisurely way for large quantities of ice to descend from mountains, and that is glaciers. These slow rivers of ice begin high up in the permanent snowfields. Snow falling in the basins between peaks builds up to great depths, squeezing out most of the air in the snow near the bottom and turning it to clear ice. Under the influence of gravity, this ice then moves slowly downhill; at about three metres a day in the case of the Franz Josef Glacier, but a more sedate 65 cm a day for the Tasman Glacier (fig. 8.5).

Glaciers end once they get low enough for the ice to melt, at a point called the terminal. In New Zealand glacier terminals have been retreating for most of this century, but in the last few years some of the West Coast glaciers have begun advancing back down their valleys. Should the Franz Josef Glacier continue to do so, it will soon extend beyond the narrow gorge where it currently ends and spill out into the broad river valley where it has not been since the 1950s. Remarkably, this is happening when most of the world's glaciers are retreating in response to global warming.

Fig. 8.6. El Niño conditions are likely to have triggered recent advances in both the Fox (above) and Franz Josef glaciers on the West Coast.

EL NIÑO AND THE GLACIERS

The cause of these advances seems to be the phenomenon known as El Niño (see Chapter 12), or the Southern Oscillation, that affects the weather around the Pacific Basin, and to a lesser extent beyond. During an El Niño, large areas of the tropical Pacific Ocean warm to several degrees above average, which raises average global temperatures by up to half a degree.

Over New Zealand, however, El Niños produce more southwest winds than normal, causing our temperatures to be *colder* than average. The reverse of El Niño, sometimes called La Niña, produces the opposite effect: average global temperatures decrease, but New Zealand gets *warmer* conditions than normal because it receives more northeast winds bringing tropical air. During the last 20 years there has been only one strong La Niña event, but there have been many El Niños, including the strongest this century which occurred over the summer of 1982–83.

The Fox Glacier (fig. 8.6), a little further south from Franz Josef, is also advancing, but interestingly the big glaciers on the eastern side of the Alps are not. There seem to be several reasons for this. Firstly, the eastern glaciers are slower to react to temperature changes and are, in fact, still adjusting to the global warming that began at the end of the last century. This is because they travel down flatter valleys and are much slower moving than the Franz Josef or Fox glaciers, and also have thick layers of gravel and stones on top which help insulate them from temperature changes.

Secondly, for most of this century, as the glaciers retreated up their valleys, they left walls of broken rock and shingle, called moraines, at their points of maximum advance.

Meltwater, trapped behind the moraines, created small lakes at their snouts, causing much faster melting than would occur if the glacier terminated on shingle or bare rock.

Franz Josef and Fox glaciers do not have moraine dams because their valleys are too steep and the shingle they produce is washed away downstream (though there was briefly a small lake in front of Franz Josef from 1939 to 1949 when the terminal was further down the valley).

There are several ways in which the colder weather of an El Niño can cause the glaciers to advance. Most important of course, is the fact that more snow falls on top of the mountains where the glaciers start. Normally, because it is brought by northwest winds, most of the rain on the West Coast of the South Island falls when temperatures are above average. In summer especially, some of the northwest airstreams are warm enough so that, even on top of the Alps, it rains rather than snows. But during El Niños these warm rains are rare, and most if not all of the precipitation on the tops falls as snow.

The Franz Josef Glacier has a very large area of snow feeding into its steep, narrow valley. Consequently, it is extremely sensitive to small climate changes and has a very fast response time of about five years between increased snow at the top of the glacier and the advance of the ice down at the terminal. Therefore, the advance beginning in 1982 may well have been triggered by the El Niño of 1977–78.

Another El Niño glacier-preserving factor is the reduction of low altitude melting. One of the best ways to melt a glacier is to surround it with warm air at close to 100% humidity; which is precisely what happens during a normal summer of northwest rains. Some of the water vapour in the warm, northwest air will then condense onto the glacial ice, releasing latent heat as it does so and causing melting. During an El Niño year, however, the rain comes with cold air from the southwest and this melting effect is much reduced.

Glaciers are subject to other influences as well. For example, in fine weather the rocks of the valley walls are heated by sunlight, and then radiate heat onto the ice causing melting. Although the ice acts like a mirror to sunlight, the heat radiated by the rocks has a much longer wavelength and is readily absorbed. Again, during El Niño summers the heating of the rock walls on the western side of the Alps is reduced because the predominant southwest winds cause more cloudy skies than normal west of the main divide.

The latest advance of the Franz Josef Glacier is still a long way short of its position last century when it was several kilometres further down the valley. But this pales in comparison with its position in the last Ice Age when it went all the way down to the sea. The sea at that time was more than 100 m below today's level and therefore 10 km west of the present coastline. Nevertheless, since a strong El Niño only finished in 1998, the Franz Josef Glacier is expected to continue advancing for a few more years yet.

CHAPTER 9

Coastal winds

The wind at sea would be a lot easier to handle if the conditions on the beach in the morning were what happened out on the water for the rest of the day. But it is not often like that. The wind changes dramatically in space as well as time. From increased friction alone, the rule of thumb is that the wind strength over the land will be half that over the sea, all other things being equal.

Most afternoons in the summer half of the year it will be windy at the beach. This is because if there is no wind being caused by the atmospheric pressure pattern, then a sea breeze is likely to develop.

SEA BREEZE

A sea breeze is a moderate to fresh wind, generally 10–20 knots (18–36 km/h), blowing from the sea onto the land, which is noticeably cooler than the conditions prevailing before it started.

It is caused by the different responses of land and water to heating by the sun. When sunlight warms the sea surface, some of the heat is conducted away to deeper water and some of the deeper, cooler water also mixes with the surface water. Consequently, the sea surface temperature does not rise much during the day. The land, however, heats up comparatively quickly – especially surfaces like concrete, asphalt and rocky terrain.

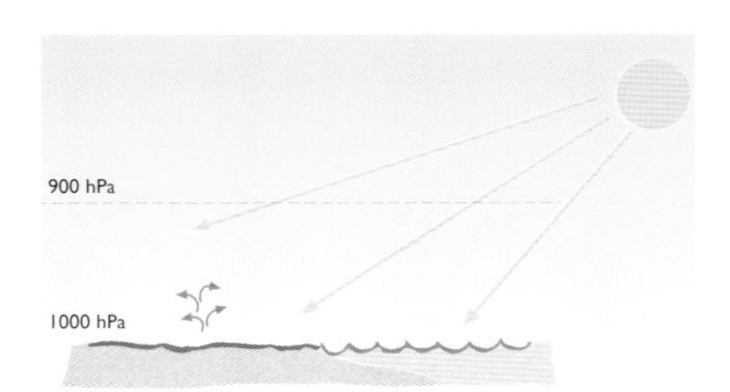

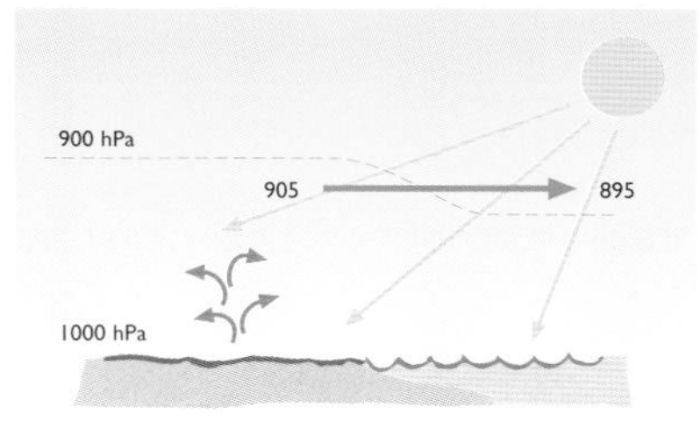

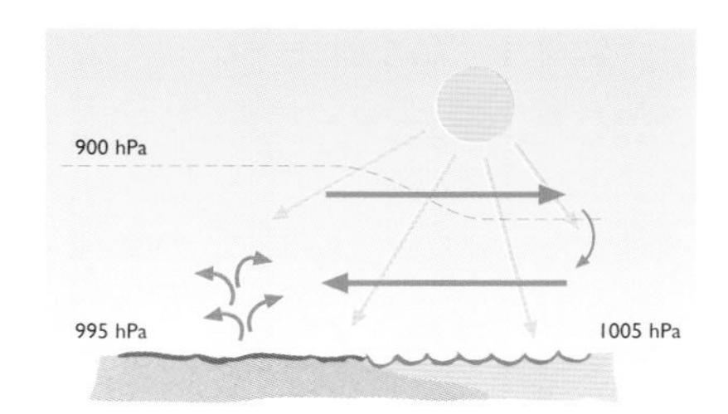

Fig. 9.1. Sea breezes. (Left) Land surface warms faster than sea surface. Air touching land heats up, rises and mixes with air above. (Middle) Temperature of layer between 1000 hPa and 900 hPa increases, causing layer to expand, lifting 900 hPa surface higher over land than sea. As a result, air about 100 m up starts to move from high to low pressure – from land to sea. (Right) Outflow around 1000 m from land to sea causes more air in vertical column above sea than in column above land. Therefore air pressure at surface rises over sea and falls over land. So, at sea level cool air begins to flow from sea over land.

Once the land heats up it warms the air immediately above it by conduction. (Daytime air temperatures can often reach 10°C higher than the overnight minimum.) The warmed air then rises and mixes with the air above. As the air in the lowest 1000 m gradually warms up, it expands. This has the effect of lifting the centre of gravity of the column of air above the land, in comparison to the column of air above the sea (fig. 9.1).

So, for example, the height at which the atmospheric pressure is 900 hPa (which is usually about 1000 m above sea level) will now be a little higher over the land than over the sea. This means that along a horizontal surface at 1000 m altitude, the pressure above the land has become a little higher than at the same height over the sea. Consequently, as air flows from high to low pressure, the air at this altitude begins to flow away from the land towards the sea.

But this means that there is gradually more air in total in the column over the sea than in the column over the land. Therefore, down at sea level, the atmospheric pressure becomes a little higher over the sea than it is over the land, so air at the surface begins to flow from the sea towards the land.

This circular movement – up over land, out to sea aloft, down over the sea, then back in over the land at sea level – is called the sea breeze circulation.

A sea breeze usually starts up late morning and drops off around dusk. Its strength depends on the temperature contrast between land and sea. Therefore, a sea breeze is often stronger in late spring/early summer than in late summer/autumn because sea temperatures are warmer in autumn than in spring and the temperature contrast between land and sea is less.

The nature of the sea breeze on any particular day also depends on what is happening with the wind before the sun begins heating the land. If there is a pre-existing wind from land to sea at about 1000 m, this generally helps the sea breeze circulation and brings the sea breeze in earlier and more strongly than would otherwise be the case. This sometimes happens when a southwest airstream affects Nelson, causing the sea breeze to come in abruptly at 25 knots (46 km/h). However, if the pre-existing wind is too strong, it will prevent the sea breeze developing at all because the air heated by the land will be quickly carried away out to sea before the sea breeze circulation can become established.

If the pre-existing wind at 1000 m is blowing in the wrong direction from sea to land, it will weaken the sea breeze or prevent its development. Also, such a wind will often bring a layer of cloud in over the land, thus reducing the amount of sunlight reaching the ground for surface heating.

Sea Breeze Direction As a sea breeze circulation develops, it reaches further out to sea and penetrates further inland, and as its length, or 'fetch', increases, it begins to be influenced by the Earth's rotation. In the Southern Hemisphere, this causes its direction to slowly swing anticlockwise. So, a westerly sea breeze blowing onto a west coast will gradually turn southwest, while an easterly sea breeze blowing onto an east coast tends northeast. Over an enclosed piece of water such as the Manukau Harbour, the sea breeze begins by blowing onto all the surrounding beaches. After a while, however, one direction usually prevails in response to the predominant distribution of land and sea. In the case of the Manukau Harbour, this is the southwest.

Sometimes two sea breezes moving inland from opposite directions will give rise to a line of thunderstorms where they meet in the middle. This happens over the Florida peninsula and also over Auckland and Northland.

WHIRLWINDS

Whirlwinds are like small cousins of tornadoes occurring in layers of the atmosphere typically one-hundredth the height of a thunderstorm. Friction will always cause the air closest to the ground to move slower than the air above it, creating horizontal rotation – as outlined in the section on tornadoes. This is most likely to become a whirlwind in places where the temperature difference between the air touching the ground and the air immediately above is greatest, such as over coastal deserts in the tropics when a cool sea breeze moves inland.

In Libya, the temperature of the air next to the ground can reach 9°C hotter than the air in the cool sea breeze about one metre up. The hot air then accelerates upwards, tilting the body of horizontally rotating air vertically, and stretching it to produce a spinning column. The resulting whirlwinds usually last just a few minutes and are between 50 m and 100 m tall.

Solander Island, Foveaux Strait – the wind at sea is often stronger than the wind over the land.

Whirlwinds become visible when they pick up dust or sand and are then sometimes known as 'dust devils'. Occasionally, they pick up water and are called 'water devils'. Similarly, these are the smaller cousins of waterspouts, but are shorter and not connected to a cloud. Usually, whirlwinds that pick up water form over land, then move over water. However, there have been accounts of whirlwinds forming over inland lakes. This is most likely after a long hot summer when the surface water of the lake has become very warm. If there is then an outbreak of cold air and a horizontal rotation pattern is present, rapid heating of the lowest air by the lake water will destabilise it, causing it to rise sharply and possibly tilt the horizontal rotation upright into a whirlwind.

A particularly vivid description of a whirlwind lifting, then dropping, a column of water over a Scottish lake in the eighteenth century has given rise to speculation that many sightings of the so-called Loch Ness Monster have, in fact, been water devils seen from a distance.

COASTAL RIVERS OF WIND

Although sea breezes can be strong enough to trouble small craft, a greater danger comes from areas of strong winds that are sometimes found just offshore, especially when an anticyclone is nearby.

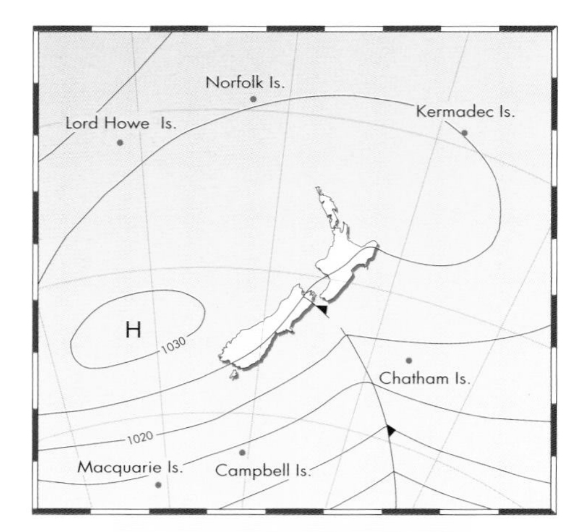

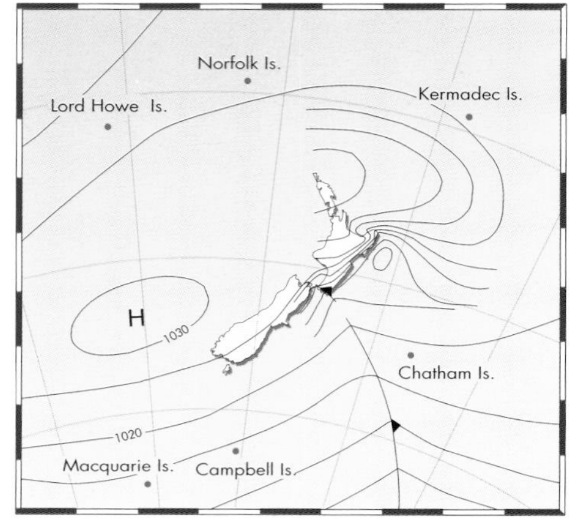

Fig. 9.2.
25 March 1992. Identical maps, except that bottom map shows isobars every 1 hPa, showing areas of strong winds not apparent at the 5 hPa spacing.

Anticyclones are commonly thought of as bringing light winds, yet they can have gale force winds relatively close to their centres. This happens because the air inside an anticyclone is generally very stable and so resists upward motion. Consequently, when the wind blowing around an anticyclone blows against a mountainous country like New Zealand and is obstructed, instead of rising over the ranges it tends to spill around the ends, much as water in a stream flows swiftly around the ends of a large rock in the middle of the current.

This creates an elongated area of strong or gale force winds extending downstream from the end of the mountain chain that is sometimes described as a river of wind. Reports from ships indicate that these rivers of wind can be hundreds of kilometres long while only tens of kilometres wide, and typically have abrupt edges where the wind speed quickly drops away.

A good example of this occurred on 25 March 1992 (fig. 9.2) when a high lay over the south Tasman Sea with a ridge extending to North Auckland. Although the isobars are a long way apart suggesting only light winds, the wind was blowing a 38 knot (70 km/h) westerly gale 50 km out to sea from East Cape. If the isobars are drawn every 1 hPa rather than every five, then the area of strong winds can be seen in the close spacing of the isobars near East Cape.

With the high and the ridge in these positions, another river of strong wind occurred to the west of the Tasman Mountains between Nelson and Buller, and extended downstream towards the coast of Taranaki, passing over the Maui gas production platform.

Strong southwest winds of around 25 knots (46 km/h) can also occur over the Manukau and Waitemata harbours when a ridge axis lies to the north of Auckland like this. The southwesterly is often strong in Auckland because the air is squeezed through the gap between the Waitakere and Hunua ranges.

The southwest wind generally brings clear skies to the Bay of Plenty and sometimes the Waikato and, during summer, the heating of the land in these cloud free areas often leads to slightly lower air pressures there, known as heat lows. When this occurs, the southwest wind can rise to a 35 knot (65 km/h) gale over the Auckland harbours because of the increased pressure difference between Auckland and the Bay of Plenty.

Another example of gale force winds near a high centre occurred on 19 April 1992 (fig. 9.3). This time the high was centred just east of Southland and Otago. A ship running down the coast of Fiordland in light winds reported an easterly gale as it moved around the corner into Foveaux Strait. At the same time, there was also a lot of wind through Cook Strait, where the southerly had just dropped below gale force at The Brothers. The strong winds extended downstream from Cook Strait to the area west of Taranaki where the Maui platform is sited.

In situations like this, strong or gale force southeast winds also occur near East Cape as the air is forced around the end of the Raukumara Range. These southeasterlies will often spread across the water offshore from Bay of Plenty and blow strongly through Colville Channel at the end of the Coromandel Peninsula, then continue flowing in a narrow river of wind downstream from the channel. Meanwhile, most of the Hauraki Gulf and the Waitemata and Manukau harbours will have winds of 15 knots (28 km/h) or less.

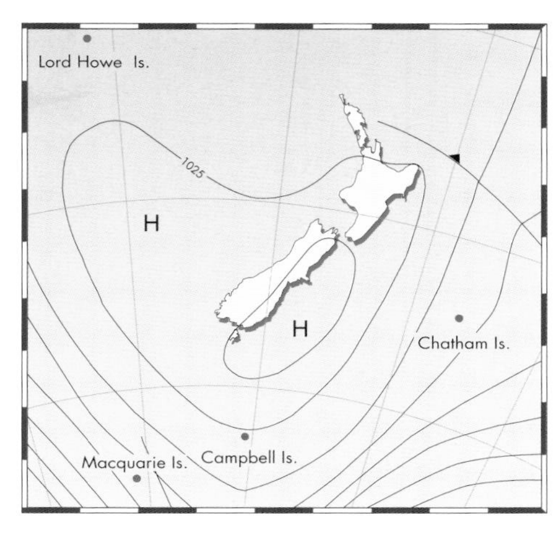

Fig. 9.3. 19 April 1992

LEE TROUGH

When the atmosphere is stable and the air is being deflected around the end of a mountain range, a trough of low pressure often occurs offshore on the downwind side of the mountains. Known as the lee trough, it can be clearly seen in the isobars east of the South Island on the map for 6 March 1992 (fig. 9.4) where northwest gales ahead of the front are blowing past the southern end of Fiordland and Stewart Island.

Although winds in the centre of a lee trough are generally light, strong winds often occur on the offshore side. On this occasion, northeasterlies of 25 knots (46 km/h) were blowing beyond the lee trough while winds at Kaikoura and Christchurch, near the centre of the trough, were around 10 knots (18 km/h).

As a front reaches the South Island in situations like this, the lee trough will often move offshore away from the land, and in so doing trigger a southerly change all the way up the east coast. In this case, the southerly was only 10–15 knots (18–28 km/h) over the land, but briefly reached 30 knots (56 km/h) over the sea.

Southerly changes of this type are often short-lived, but are frequently followed within 12 hours or so by a more sustained and stronger southerly as the front moves east and the next high comes on. This sort of behaviour by a lee trough makes forecasting southerly changes a lot more difficult.

DOWNSLOPE WINDS

Aside from contributing to lee troughs, the hills can also dramatically affect the wind over the sea in another way. If the wind begins to blow strongly around the end of mountains or through a gap like Cook Strait, some of the air may end up being pushed up and over the hills even though the atmosphere is stable. When this happens, the air accelerates on the downslope side of the hills and reaches very high speeds as it returns to sea level.

Cook Strait in gale force winds.

Downslope winds tend to occur in favoured locations – like Karori Rock on the Wellington south coast. When 40 knot (74 km/h) northerly gales are blowing through Cook Strait, steady winds of around 70 knots (130 km/h) are often reported by the Cook Strait ferries near Karori Rock as the wind rises over the southern hills and accelerates down the other side.

The same thing often happens in parts of Palliser Bay; fishing boats kilometres offshore have been known to be hit by stones picked up by the wind over the land. On 12 March 1992 (fig. 9.5) a ship approaching Wellington Harbour from Palliser Bay in a 35 knot (65 km/h)

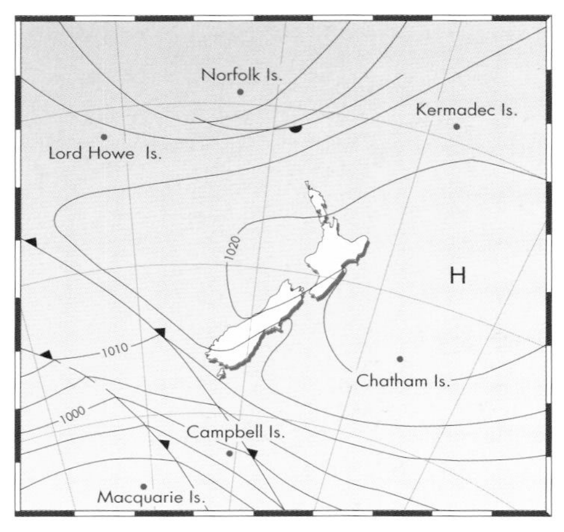

Fig. 9.4. 6 March 1992

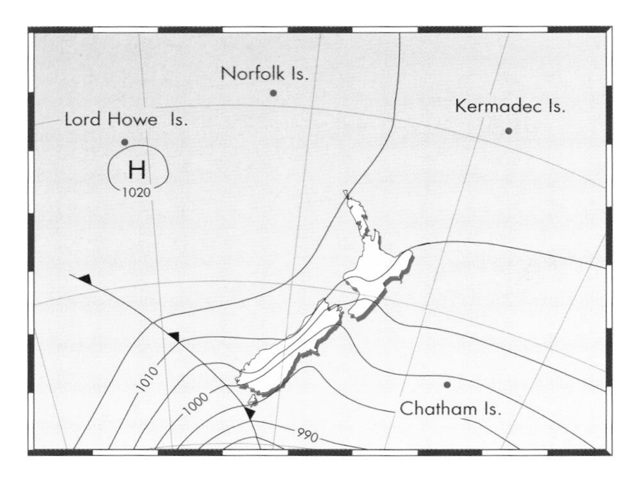

Fig. 9.5. 12 March 1992

gale experienced a jump in wind strength to 57 knots (106 km/h) when it was about 25 km south of the southern tip of the Rimutaka Range.

Castlepoint, on the Wairarapa coast, is also affected by very strong northwesterlies, sometimes recording 60 knots (111 km/h) gusting 80 knots (148 km/h) when Cook Strait may only be experiencing around 35 knots (65 km/h). This often happens in conjunction with a lee trough which, as it moves away from the coast, can produce a brief southerly change similar to the one described off the Canterbury coast.

Because these gales are blowing from land to sea, the waves they produce are not especially large. They can, however, still prove hazardous to small craft as the wave crests are very close together and become progressively larger the further offshore they get.

Captain Cook

CAPTAIN COOK'S DEATH

Sudden strong squalls caused by hills or mountains interacting with the wind occur all around the world and one of these played a role in the events leading up to Captain Cook's death in 1779. Several days after leaving Kealakekua Bay on the island of Hawaii, Cook sailed the *Resolution* into an area of violent wind squalls caused by the volcanic peaks of Hawaii disrupting the trade winds. The *Resolution* was so badly damaged that Cook was compelled to return to the bay for repairs.

This he was reluctant to do, feeling that he had already outstayed his welcome. Although Cook seems to have been treated almost as a god, the visit of his two ships had imposed a considerable strain on the local food supply. On returning, he and his crew were treated with far less respect than before. Theft increased dramatically, culminating in a cutter being stolen from its moorings overnight. Cook then attempted to kidnap a chief to force the return of the boat. This led to the fight on the beach in which Cook, four marines and 17 Hawaiians (four of them chiefs) were killed.

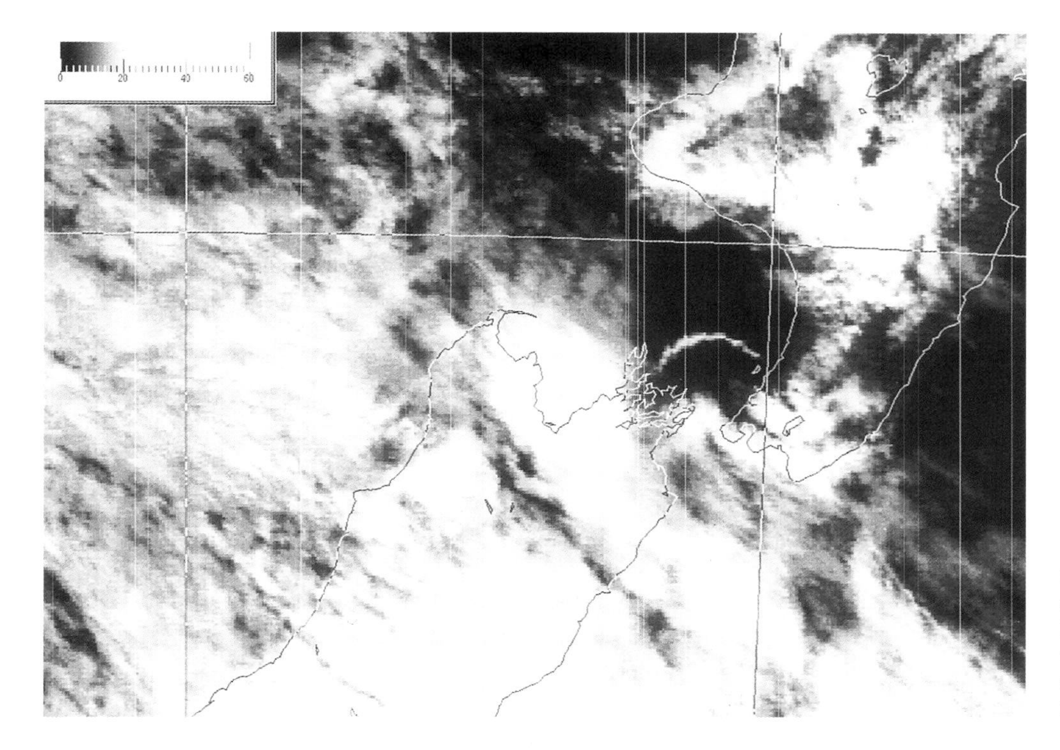

Satellite photo showing an arc of cloud forming at the leading edge of a southerly change bursting through Cook Strait on 26 August 1993.

CURVATURE OF THE ISOBARS

The nature of southerly changes moving up the Canterbury coast to Cook Strait depends on the curvature of the isobars. In Chapter 5 we saw the case of 18 October 1994 (fig. 5.7) when the isobars were cyclonically curved around a low east of the South Island. On that occasion, the southerly wind change was much stronger at Kaikoura than in Cook Strait because the air flow was parallel to the South Island and concentrated on the Kaikoura coast by the mountains' proximity to the sea.

The opposite situation occurred on 4 September 1993 (fig. 9.6) when the isobars were anticyclonically curved where they impinged on the South Island and the southerly wind was much stronger through Cook Strait than along the Canterbury coast. In this instance, the wind at The Brothers in Cook Strait reached 64 knots (120 km/h), but at Kaikoura only reached 30 knots (55 km/h) because much of the wind blowing onto the Kaikoura mountains was channelled northwards and poured around the end into Cook Strait. The storm force winds (storm force being mean wind between 48 and 63 knots or 88 and 117 km/h, whereas a gale is mean wind between 34 and 47 knots or 63 and 87 km/h) blew for a day and a half through Cook Strait, with The Brothers recording 11 hours above 60 knots (111 km/h) followed by a further 24 hours above 50 knots (93 km/h).

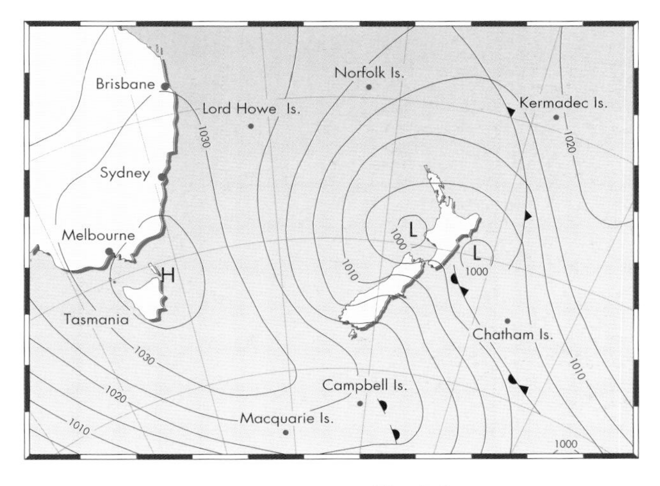

Fig. 9.6.
4 September 1993

In these sorts of southerlies, the wind through Cook Strait does not typically reach its maximum strength straight away, but rather builds up over 12 hours or more after the onset of the southerly.

FAST CHANGING SITUATIONS

When going to sea in a small boat it is important to get the latest forecast before setting out because not only can the weather change quickly, but often the forecast will too, as new information becomes available. A good example of this occurred on Wednesday 22 November 1989 (fig. 9.7) when a low deepened rapidly as it moved unexpectedly towards Auckland from the northeast. At midday the wind over most of the North Island coastal waters was around 20 knots (37 km/h), but by midnight it had risen to a 35 knot (65 km/h) gale in many places as the difference in pressure across the North Island increased.

Through Cook Strait the wind was over 50 knots (93 km/h) for more than 12 hours, making it impossible to search for two yachtsmen whose catamaran had capsized near

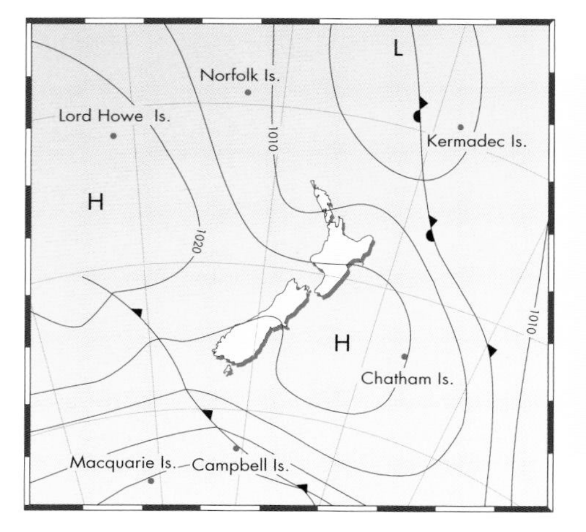

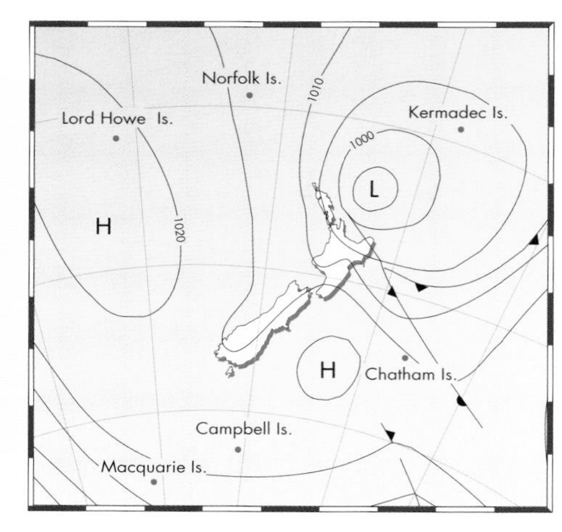

Fig. 9.7. Midnight, 21 November 1989 Midnight, 22 November 1989

Kapiti Island just before dusk. On Thursday morning, winds of 70 knots (130 km/h) were recorded at the Maui platform off the Taranaki coast. Once the low crossed the land, however, it weakened rapidly and the wind speed at Maui halved within six hours.

In view of the dangers posed by Cook Strait, it is interesting to note the deep respect that Maori had for this stretch of water – as illustrated in a story told by Tamihana Te Rauparaha in his book about his father *History of Te Rauparaha*. When someone made their first crossing of Raukawa, as Cook Strait was known, a blindfold was tied over their eyes and they paddled the whole way across without seeing their surroundings. On reaching the far shore, they were carried to land without touching the water. Then the karaka leaves covering the prow of the canoe and the blindfold would be taken and left at a sacred place, in gratitude for a safe crossing.

CHAPTER 10

Convergence lines

When the pattern of low-level winds drives two masses of air towards each other, upward air motion occurs in the convergence zone where they meet, forming cloud, and often rain. If the two air masses have markedly different temperatures, the convergence zone will become a front. On other occasions there will be little temperature difference between them, and it is this type of convergence zone that often occurs inside anticyclones.

CONVERGENCE IN AN ANTICYCLONE

When air descending in an anticyclone has made the atmosphere very stable, it resists upward motion and tends to flow around obstacles like the South Island, rather than over them. In such instances, an airstream blowing onto the South Island then splits into two branches. Sometimes these wrap around the South Island and come together again on the downwind side creating a convergence zone, in much the same way as a wave rushing up a beach will part to go around a rock, then charge into itself from opposite directions. The convergence zone will usually create enough upward air motion to form cloud, and sometimes even rain, despite the stable conditions caused by the anticyclone.

An example of this occurred on 10 May 1995 (fig. 10.1) when two airstreams separated by the South Island came back together northwest of Nelson forming a line of cloud with showers. More than 24 hours after forming, this cloud crossed over Wellington bringing a brief period of light rain that temporarily disrupted air traffic and marred an otherwise fine day.

Little rain if any fell from the line of cloud as it crossed Wanganui and Taranaki later in the day, but it produced 6 mm on the Waikato coast the next afternoon, as well as a prolonged period of fog and drizzle that closed Hamilton Airport and a period of rain in Auckland in the evening.

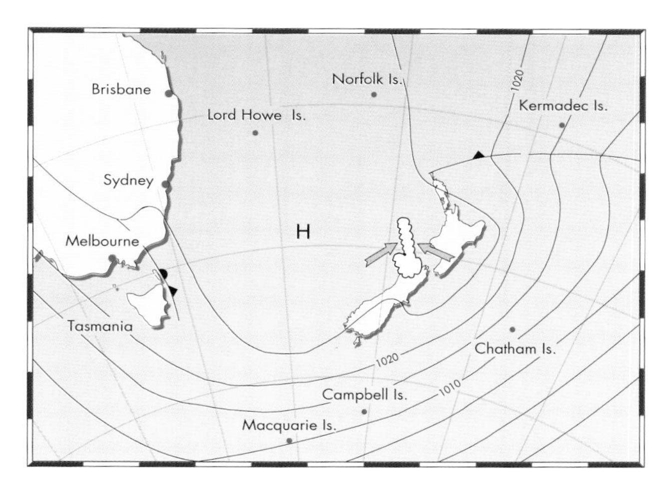

Fig. 10.1. 10 May 1995. Two airstreams converge, causing rain in Wellington.

The line was not a front because the air either side of it was of similar temperature and humidity, being from the same source and having changed little during its two different journeys around the South Island. Convergence lines like this are not marked on the weather maps that appear on television or in newspapers as they are too small, usually too short-lived, and often do not produce rain.

Because this type of convergence occurs within a stable atmosphere, the cloud that develops is usually shallow – between 1000 and 2000 m thick. This is normally too shallow for rain to develop; any cloud droplets starting to grow towards raindrop size tend to fall out of the cloud too quickly, while still too small, and evaporate long before reaching the ground.

Also, the cloud is low in the atmosphere, with the base typically less than 1000 m above sea level. Consequently, the temperature inside the cloud is not cold enough to allow ice to form and help facilitate raindrop development.

The likely explanation for the rain that does fall on occasions is that the convergence somehow produces continuous upward air motion inside the cloud, rather than the intermittent bursts of updraughts typical in most cumulus clouds. Continuous upward air motion could enable water droplets to grow to raindrop size over many hours by keeping them suspended in air that was rising at about the same speed as they were falling.

Once the droplets reached raindrop size, the pull of gravity would be sufficient to make them fall through the air faster than it was rising, and so progress down out of the cloud.

Convergence lines of this type, forming northwest of Nelson, have a habit of turning up in Wellington with an hour of light rain on the first day of a northerly. That this often tends to happen mid to late morning, suggests there is some other factor contributing to their development that has to do with the regular day/night variations of heat and cold.

The most likely explanation is that, as the convergence line moves across the sea north of Nelson, it runs into cold air that has developed over Wanganui and Manawatu on a calm, clear night, then slid downhill as a katabatic wind and flooded out over the water. Although this cold air is in a very shallow layer, the concave curve of the coastline south of Taranaki means a relatively large amount of air could move over a relatively small area of ocean.

Once the convergence line ran into the colder, denser air, the colder, denser air would slide underneath, pushing the convergence line up and increasing the chance of rain. The air from the convergence zone would also be forced to rise as it moved over the low hills around Wellington, further increasing the likelihood of rain. In the example, there was 2 mm of rain in the hill suburbs of Karori and Wadestown, but only 0.4 mm at Wellington airport.

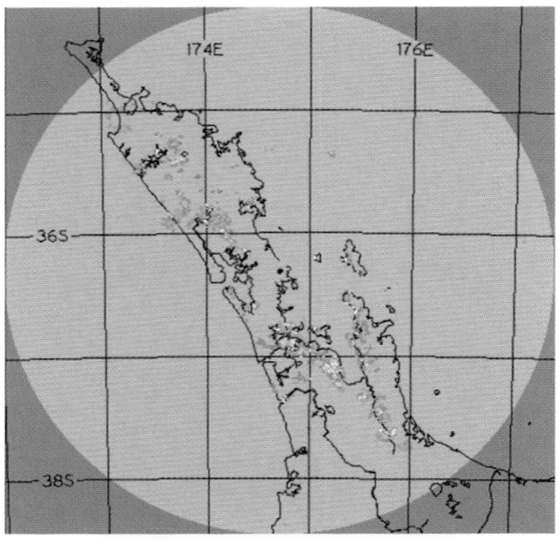

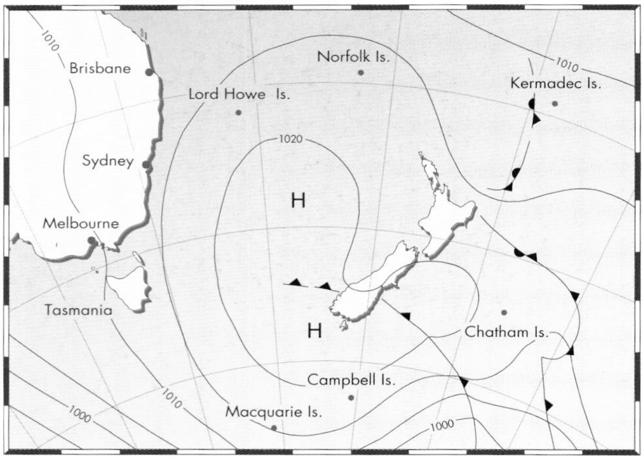

Fig. 10.2.
19 January 1996. The radar shows rain falling from cloud formed by convergence of sea breezes from east and west coasts.

SEA BREEZE CONVERGENCE

When a sea breeze moves inland, cloud will often form above its leading edge due to a combination of convergence and upward air motion from the heating of the land (which would have given rise to the sea breeze in the first place). These clouds will sometimes produce showers if the air aloft is cool enough to allow the warmer air to rise up through it and form deep cumulus clouds.

A favoured location for this to happen is the western edge of the Canterbury Plains where the foothills start. In the summer half of the year, the temperature at the edge of the foothills has usually become quite high by the time the sea breeze has worked its way across the plains. In addition, the air also gets extra uplift when it hits the rising ground. The resulting showers usually form a broken line along the edge of the foothills and although they can be heavy, gradually die out in the evening as the land cools.

When the configuration of the land is such that two sea breezes blow towards each other from opposite directions, the convergence zone where they meet will often produce showers. This can happen in a number of places around New Zealand. For example, in inland Taranaki where the sea breezes from the north and south coasts come together in the late afternoon in a line stretching east from Mount Taranaki. Also, in the Raukumara Range when the sea breeze from Gisborne meets the sea breeze from eastern Bay of Plenty, and in the Tasman Mountains of northwest Nelson where the sea breezes from Buller and Golden Bay blow towards each other from opposite sides of the ranges.

In Christchurch, the developing northeast sea breeze sometimes runs into a southerly sea breeze coming up from Lake Ellesmere or a more general southerly that is weakening and dying out. Any showers produced as a result will tend to be in the Cashmere to Addington region.

In Wellington, a developing southerly sea breeze sometimes converges with a light northerly and produces a very black line of threatening cloud over the city. This usually produces no rain and vanishes within an hour, but occasionally it will sprinkle a few drops for five minutes or so.

In Auckland, however, the two sea breezes approaching from opposite coasts can sometimes create a convergence zone that gives rise to slow-moving cumulonimbus clouds. These can bring heavy rain to parts of the city while leaving other areas dry. This occurred on 19 January 1996 when over 50 mm of rain fell in some parts of the city causing flash flooding, disrupting traffic and ruining a large outdoor concert at Ericsson Stadium (fig. 10.2).

A singularly dramatic example of sea breeze convergence occurred on 14 February 1988 when over 200 mm of rain fell in four hours over a small area between Coatesville and Paremoremo to the north of Auckland city.

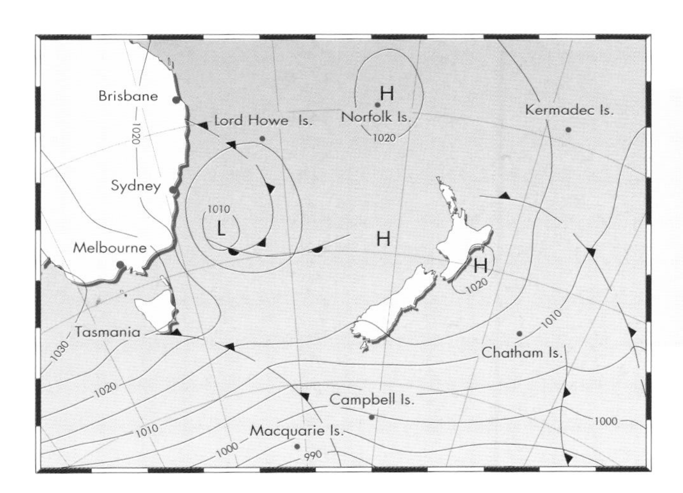

Fig. 10.3. Easterly convergence line approaching Auckland, 12 May 1996.

CONVERGENCE ZONE FROM THE EAST

Another convergence line that sometimes brings showers to Auckland comes from the east, though it is initially caused by a southerly change spreading up the east coast of the North Island and bursting past East Cape. Part of the southerly air then spills around into the Bay of Plenty as an easterly. A line of cloud usually forms along the leading edge of this wind change and can be readily seen on satellite pictures swinging round onto the land (fig. 10.3). On occasion, this scenario produces heavy rain in Whakatane and other parts of the Bay of Plenty. In Auckland, the showers are usually brief and not especially heavy, and are often confined to the Eastern Bays, Great Barrier Island and the Coromandel Peninsula.

WELLINGTON/HUTT VALLEY FLOOD 20 DECEMBER 1976

Extremely heavy rain fell on 20 December 1976 from a convergence zone that ran along the hills west of Wellington city and stretched up over the Hutt Valley. More than 300 mm was recorded in 24 hours over much of this area – something expected, on average, only once every 100 years or more. Much of the rain actually fell in less than 12 hours, making it more like a once-in-700-years event.

Flash floods metres high roared down steep gullies where, ordinarily, ankle-deep creeks trickled into the harbour. Miraculously, no one was killed by these, despite the Hutt motorway being cut and hundreds of workers having to be rescued by helicopter from the roof of a factory surrounded by fast moving flood waters from the Korokoro

Fig. 10.4. A fireman leads a family from their home during floods that struck Hutt Valley in 1976.

stream. Many vehicles were destroyed, some crushed almost beyond recognition.

The torrential rain caused landslips in many places, tragically taking the life of a three year old boy when the side of a hall in Crofton Downs collapsed under a slide of rocks and earth. A number of houses in Stokes Valley were crushed by slips or driven from their foundations. In the Hutt Valley, a state of emergency was declared with many people having to be evacuated. After the flood waters receded, damage was estimated at $30 million (fig. 10.4).

The causes of this convergence zone can be understood by looking at the large scale situation. The weather map for 20 December 1976 (fig. 10.5) shows a low east of Cook Strait with a shallow trough of low pressure extending across Wellington to another low northwest of Auckland. The second isobar around the low bends across the South Island from Kaikoura to Hokitika, then loops back onto the North Island near Wanganui. The large gap between it and the central isobar of the low where they both bracket the Wellington region, indicates very light winds for the Wellington area.

The mountain ranges block most airstreams crossing the country so that through the narrow gap of Cook Strait, the wind is almost always funnelled directly from high to low pressure rather than blowing parallel to the isobars. On this occasion, a light southerly was blowing along the Kaikoura coast towards the middle of the trough over Wellington while, at the same time, a light northerly was blowing down the Kapiti coast towards Cook Strait. These two low-level winds, blowing from opposite directions, met in a convergence zone over the Hutt Valley and the hills west of Wellington city.

The air in the trough of low pressure was very unstable and so deep cumulonimbus shower clouds were already mushrooming up. Once the convergence zone was established, the upward air motion driving the cumulonimbus clouds increased dramatically, and heavy rain commenced.

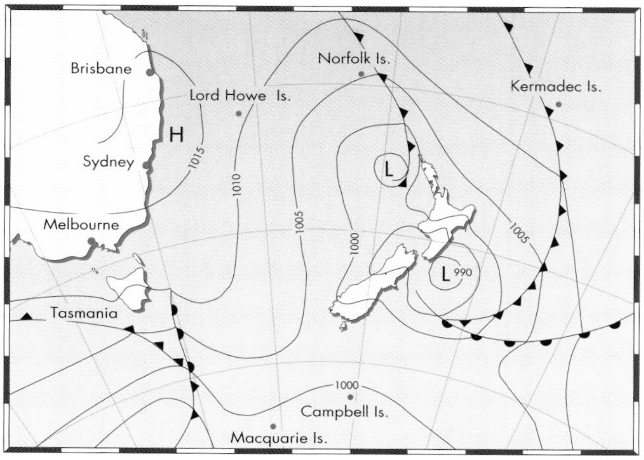

Fig. 10.5. Wellington/ Hutt Valley flood, 20 December 1976.

The upward air motion within the clouds helped to sustain and reinforce the inflow of air at the base – thereby setting up a feedback mechanism between the clouds and the convergence zone which served to maintain and intensify them both. Furthermore, the way in which the heaviest rain followed the line of the hills suggests that the cumulonimbus became anchored over them for a time, perhaps because the surface northerlies on one side and the southerlies on the other were partially deflected upwards by the rising ground, helping to focus the upward air motion near the ridge line.

Eventually, the trough over Wellington moved slowly to the east allowing the southerly winds to prevail, and the convergence line broke up. The cumulonimbus began to dissipate and the heavy rain eased off.

SQUALL LINE CROSSES AUCKLAND

Although low-level convergence happens frequently, it often results in not much more than lines of shallow cumulus with little or no rain. To get more drama out of a convergence zone, some other mechanism is necessary to help drive the air upward.

One way this happens is when a fast moving trough in the upper atmosphere catches up with a shallow convergence zone in the lower atmosphere. Upward motion occurs ahead of the upper trough, which then cools the air in the middle atmosphere by expansion, thus creating an environment through which cumulus clouds from the convergence zone lower down can grow explosively.

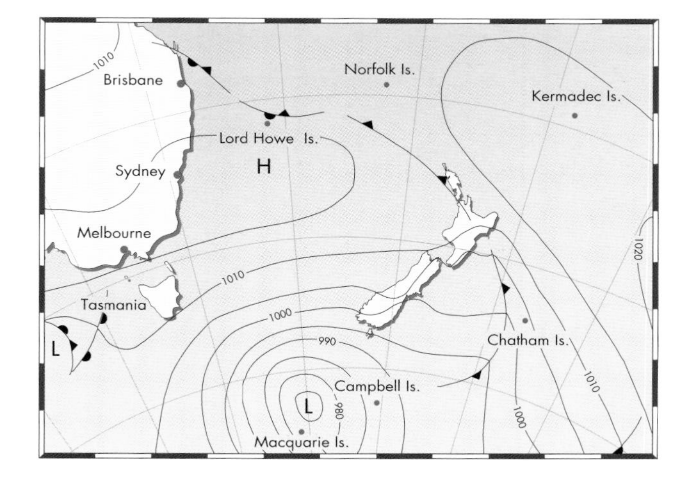

Fig. 10.6. Auckland squall line, 4 February 1996.

This happened on Sunday 4 February 1996 when an upper trough moving rapidly across the Tasman Sea, caught up with a line of shallow cumulus just west of Auckland. The rising, expanding air ahead of the upper trough then cooled the middle atmosphere by about 5°C and, in the space of a couple of hours, the shallow cumulus cloud less than 2000 m deep grew into towering cumulonimbus about 10,000 m tall. As the line crossed the city, there was a brief period of very heavy rain in a number of places as well as squalls and thunderstorms (fig. 10.7).

By around midday the upper trough had caught up with the convergence line and was in the process of overtaking it. As this happened, the thunderstorm activity dissipated. However, there was now a broad band of stratiform cloud in the middle and high atmosphere, formed from water vapour that had been lifted from near sea level by the thunderstorms, and light rain fell from this over a broad area.

On the 6.00 a.m. weather map (fig 10.6, previous page), the line of thunderstorms was marked as a cold front lying just west of Auckland. However, just six hours earlier at midnight, there was no cold front in the area, the nearest being some 400 km away in the Tasman Sea. This distant front was, in fact, a separate cloud band that had been created

Fig. 10.7. Radar images shows rapid intensification and dissipation of showers in a squall line that formed west of Auckland on 4 February 1996.

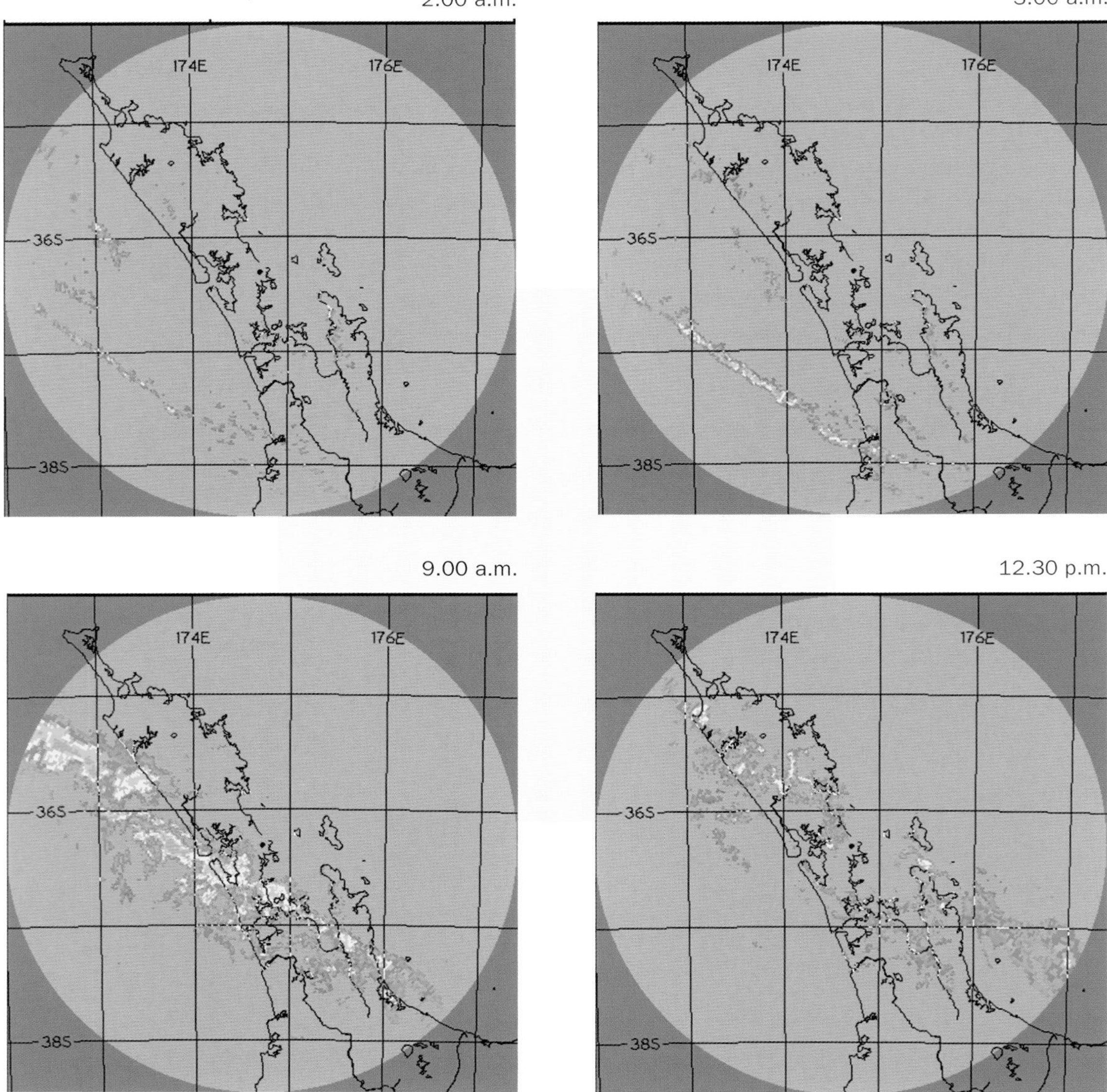

by the upper trough during the previous 12 hours, and which, at midnight, it was just starting to overtake and abandon to decay.

This process is not uncommon and creates the impression of a front jumping unrealistic distances when there are actually two different cloud bands, each created and then abandoned in turn by the upper trough.

SOUTH CANTERBURY FLOOD

Fronts, as we have seen, are also examples of convergence zones; ones where two bodies of air of very different temperatures and densities come together.

The strength of the convergence can vary considerably during the lifetime of the front and relates to the characteristics of the converging air. The larger the difference in temperature, and therefore origin, of the air, the stronger the uplift associated with the front and the heavier the rain.

On 12 March 1986, air from the tropics met air from the subantarctic over Canterbury. The warm, moist air from the tropics had been started on its journey towards New Zealand four days earlier by a tropical cyclone well to the north of the country. It was then propelled the rest of the way by a northeast airstream flowing between a large, slow moving anticyclone east of the Chatham Islands and a low over the Tasman Sea (fig. 10.8). The cold air from the south was pushed up over Canterbury by a rapidly intensifying anticyclone moving across the south Tasman Sea from near Tasmania. At midday, the southerly air over Dunedin had a temperature of only 13°C compared to the 21°C that the northeast flow was bringing to Christchurch.

For a time, the forces pushing the cold and warm air towards each other were about equal and the frontal convergence zone was held stationary over Canterbury. While this lasted, the warm air rose over the cold causing heavy rain and thunderstorms.

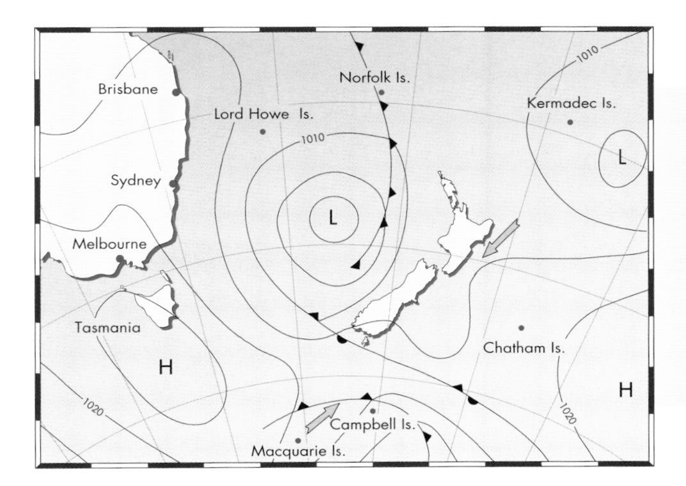

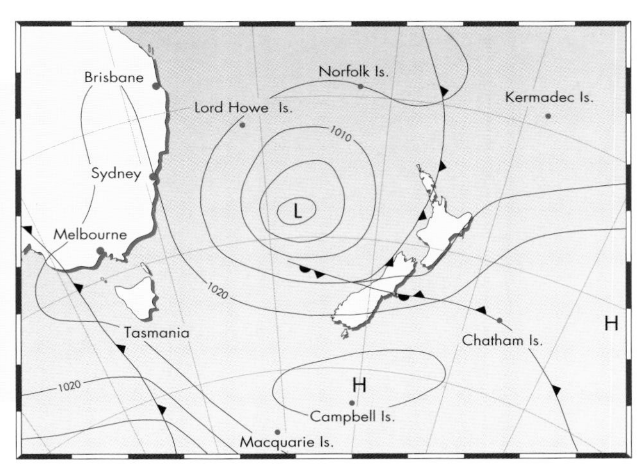

Fig. 10.8. Floods in Canterbury 12 March 1986 13 March 1986

Cumulonimbus clouds grew to 13,000 m (as judged by recorded cloud-top temperatures of minus 70°C).

In mid and south Canterbury, rain in excess of 250 mm was recorded, much of it in less than 12 hours. Rainfalls in some places were greater than those calculated to occur, on average, once every 400 years. In addition, two to three times normal rainfall had occurred over much of the area in the month preceding the downpour, so the soil was already nearly saturated and unable to absorb much more rain.

The floods caused an estimated $60 million worth of damage. The Tengawai River broke its banks, drowning one farmer who was swept from his tractor, and inundating the town of Pleasant Point, all of whose 1200 inhabitants had to be evacuated. Many homes in Temuka were also evacuated, as well as all the farms in the Hakataramea valley – which had just suffered several years of drought. Flooding also occurred in Christchurch and Rangiora.

The storm ceased abruptly because the anticyclone lying south of New Zealand kept moving east, and in doing so changed the direction of the cold air from southeast to northeast. Now the warm and cold air were no longer converging, but instead moving in the same direction.

This is a good example of how a flood, rather than having a single cause, was the combined result of two anticyclones, a depression and a tropical cyclone, which delivered two very different lots of air to the same corner of New Zealand at the same time.

CHAPTER 11

Ex-tropical cyclones

The most intense tropical cyclones, known as hurricanes in the North Atlantic and typhoons in the North Pacific, cause more destruction than any other type of weather event. In August 1992, hurricane Andrew inflicted damage totalling over US$30 billion on Florida and Louisiana, causing seven insurance companies to fail and sending shock waves through the insurance industry worldwide. Had Andrew made landfall just 100 km further north, it is estimated that the damage bill could have been as high as US$100 billion. The death toll was 53 but would have been higher had not accurate forecasting allowed for the timely evacuation of more than a million people from Andrew's path.

Tropical cyclones have a long history of causing havoc to shipping. In 1281 when Kublai Khan, the Mongol Emperor of China, sent his armies to invade Japan, his fleet was destroyed by a typhoon and the invasion failed. The Japanese called this storm Kamikaze, or 'the divine wind'.

Less than two years after the death of Captain Cook in 1779, his 15 year old son Nathaniel was drowned when his ship, *Thunderer*, went down with all hands off the coast of Jamaica in a hurricane that sank 13 Royal Navy ships.

One of the more unusual consequences of a tropical cyclone occurred on the small island of Pingelap in the North Pacific, which is nowhere more than three metres above sea level. In 1775, the high seas and enormous waves associated with a typhoon swept over the entire island killing around 90% of the population. The island was scoured clean of vegetation, including coconut palms, and most of the survivors subsequently died of starvation. In all, the population was reduced from around 1000 to about 20. Numbers gradually built up again, but today the population has an extremely high incidence of colour blindness – 1 in 12 as compared to less than 1 in 30,000 elsewhere in the world. Apparently the hereditary king, who survived the typhoon, was a carrier of the gene responsible for colour blindness.

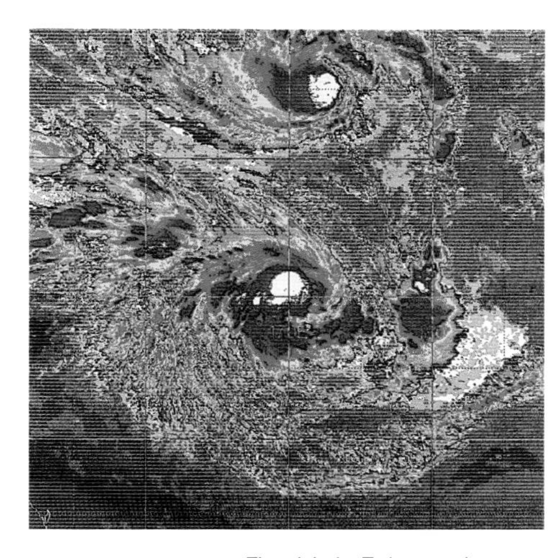

Fig. 11.1. Enhanced satellite image of a tropical cyclone pair that formed on either side of the equator in January 1991. Because the circulation in the Northern Hemisphere is opposite to that in the Southern Hemisphere, the northern cyclone Axel, is a mirror image of Betsy in the Southern Hemisphere.

WHAT ARE TROPICAL CYCLONES?

Tropical cyclones are revolving storms about half the size of the mid-latitude depressions we experience around New Zealand, but with a pressure gradient about ten times stronger. If you drew all the isobars around a tropical cyclone on a standard weather map, they would be so close together near the centre they would touch, leaving just a blob of black ink on the page representing a core of extreme low pressure.

Consequently, huge falls in pressure occur as tropical cyclones approach, followed by

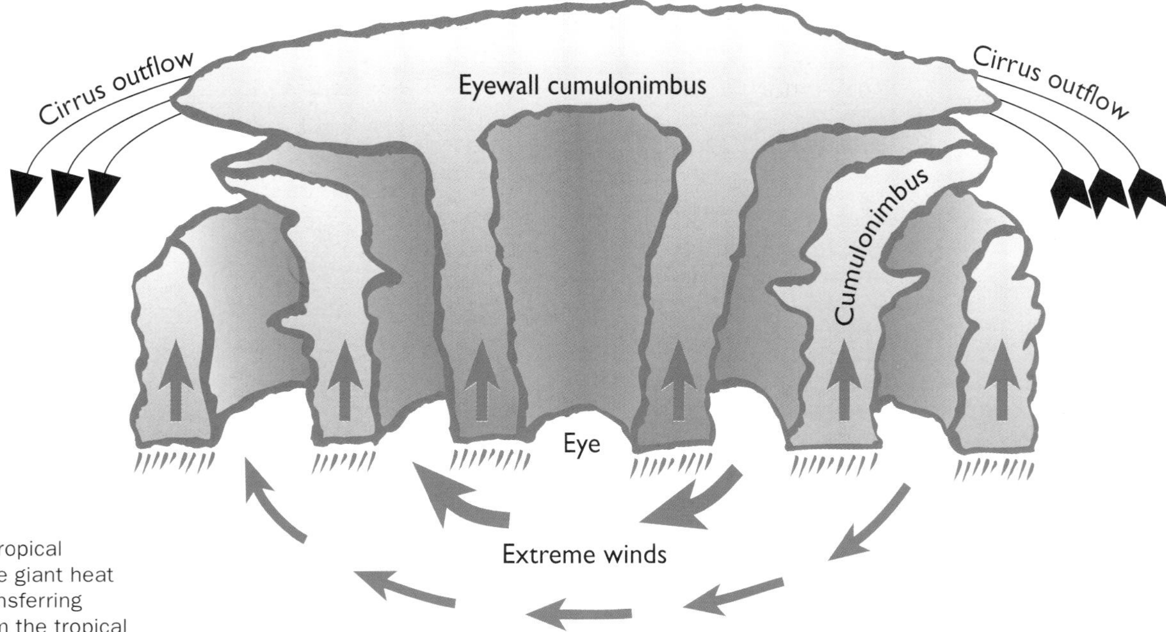

Fig. 11.2. Tropical cyclones are giant heat engines transferring warmth from the tropical oceans into storms with ferocious winds and torrential rain. The heat released in a major tropical cyclone in one day is the equivalent of all the electricity used in the United States in about six months. Although these storms cease to be tropical cyclones as they move towards New Zealand they still produce some of the most destructive weather over the country.

dramatic rises as they move away. For example, when typhoon Oscar passed over the small island of Hachijojima off the coast of Japan in September 1995, the pressure rose by 55.6 hPa in the space of three hours as it moved away. Sea level pressures in the centres of tropical cyclones can get extremely low: a pressure of 870 hPa was recorded near Guam in the northwest Pacific as typhoon Tip passed by in October 1979. In New Zealand a pressure this low would normally be found half way up Mount Cook, or near the tops of the Tararua and Ruahine ranges.

Tropical cyclones (fig. 11.2) have a central core of warm air containing a narrow belt of extreme winds which encircle a small area where the winds are much lighter and the skies may even be clear. This is known as the eye of the cyclone. The eye is separated from the belt of extreme winds by a ring of very tall, heavy-rain-producing cumulonimbus clouds.

This ring of clouds is known as the eyewall. When a tropical cyclone intensifies, which often happens in bursts or pulses lasting about 24 hours, the ring of clouds comprising the eyewall contracts into the eye. As it narrows, another ring forms further out and eventually replaces the first eyewall, which dissipates. During this replacement, which takes about 12 hours, the cyclone usually weakens a little.

Tropical cyclone ingredients and where to find them Tropical cyclones form over warm tropical oceans, usually in the summer half of the year, although in the northwest Pacific they can form at any time.

In the South Pacific, the tropical cyclone season is from November to April, though occasionally one forms a month or two outside this range. The average is about nine per season, but there have been as many as 17. Tropical cyclones don't form over the South Atlantic or the part of the South Pacific near South America because of the cold ocean currents.

Sea surface temperatures higher than about 26.5°C are necessary to supply the heat needed to drive tropical cyclones. Heat is provided to the air directly, by conduction at the sea surface, as well as indirectly at higher levels as the water vapour evaporated from the sea releases latent heat when it condenses back to liquid. The air gains about four times the heat energy from the condensing water than from the direct heat transfer at the sea surface.

The seeds that give rise to tropical cyclones are clusters of very active thunderstorms that, over the Pacific Ocean, are usually found in broad bands along the Inter-Tropical Convergence Zone or the South Pacific Convergence Zone. These are places where the trade winds converge, and these winds also provide some of the spin or vorticity needed for tropical cyclone growth.

Cyclones must rotate in order to grow, and they derive some of this rotation from the spin of the Earth. Stationary air is not in fact stationary when viewed from space because it is always rotating one revolution per day with the Earth. Because the Earth is a sphere, the speed of rotation depends whereabouts on the Earth's surface you are. At the equator you would be moving at approximately 1670 km/h, whereas if you moved towards the poles you would get closer to the Earth's axis of rotation and consequently the speed of rotation would decrease. Air at the latitude of Auckland, for example, is rotating through space at about 1330 km/h, while air over Invercargill is only rotating at approximately 1150 km/h. (So there is truth in the rumour that life moves at a slower pace south of the Bombay Hills!)

When air from different latitudes is drawn into a cyclone, the difference in rotational speeds around the Earth's axis contributes to the rotation of the cyclone. Air moving away from the equator towards a developing low will be rotating faster than the Earth's surface beneath the low, and this adds a westerly component to the air's movement. Air moving away from the pole towards a developing low will be rotating slower than the Earth beneath the low, and this adds an easterly component to the air's movement.

However, this effect only operates at latitudes greater than about four degrees from the equator. Closer than this and the difference in rotation speed per degree in latitude is too small to have an appreciable effect. Therefore tropical cyclones are unable to form closer than four degrees to the equator (fig. 11.1).

The cumulonimbus thunderstorm clouds that start off a tropical cyclone grow best when there is cool air at high altitudes associated with a trough in the upper atmosphere. But the cumulonimbus carry large amounts of heat aloft, ultimately replacing the high level trough with an anticyclone in the upper atmosphere. Air at high levels then flows away from the tropical cyclone forming the characteristic disc of cloud hundreds of kilometres in diameter readily seen on satellite pictures. This outflow of air aloft further lowers the sea level pressure of the tropical cyclone, thereby increasing the inflow of warm moist air feeding into the base of the cumulonimbus clouds – yet another example of a feedback mechanism working in the atmosphere.

Migration Once formed, tropical cyclones gradually move away from the tropics (fig. 11.3). In the central South Pacific they often move southeastwards, but those that form in the western South Pacific and closer to the equator usually move southwest to begin with, then may recurve and move southeast. Picking the time and place of this change of direction is a major forecasting challenge.

Because a tropical cyclone's area of damaging winds is relatively small and the Pacific Islands sparsely scattered in the ocean, most tropical cyclones pass harmlessly between islands. If, however, a tropical cyclone passes over an island, the damage can be catastrophic. The worst scenario is when a tropical cyclone becomes slow moving next to an

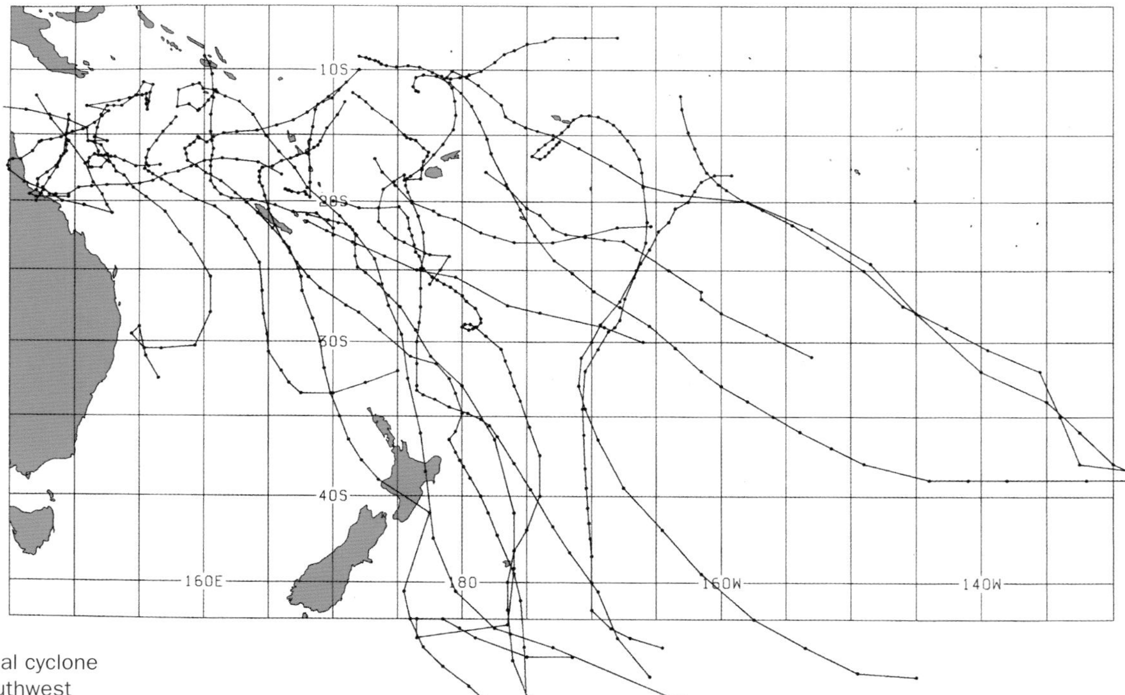

Fig. 11.3. Tropical cyclone tracks in the southwest Pacific between February 1993 and June 1997, including Drena (which crossed from Taranaki to Wairarapa) and Fergus (Bay of Plenty to Hawkes Bay).

island, as tropical cyclone Val did over Samoa in December 1991 (fig. 11.4).

The wind speeds associated with Val were estimated at up to 170 km/h, with gusts to 240 km/h. Twelve people were killed and damage was estimated at $535 million. On Savai'i, hardest hit of the two larger islands, 95% of the houses were reported wrecked and 90% of the foliage stripped from the trees. In some coastal areas the sea was driven onto the land by a combination of the extreme winds and the intensity of the low pressure, which was estimated to be 935 hPa at the cyclone centre.

Problems did not end once Val moved away. Lack of food, clean water and shelter caused disease and illness. Longer term problems included the pollination of plants: a task normally accomplished by fruit bats, more than 90% were thought to have been killed or blown away.

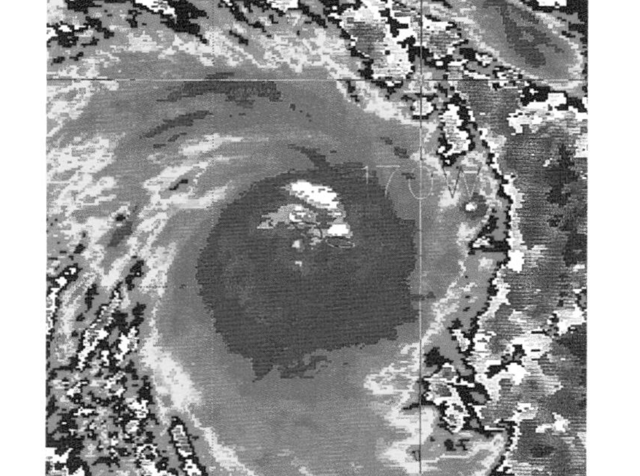

Fig. 11.4. Tropical cyclone Val, 8 December 1991. Colours correspond to temperature: white patches are -90˚C. Blue is next coolest, then red, yellow, green and black.

Transition to mid-latitude depression Tropical cyclones mostly accelerate as they move away from the tropics, but occasionally, if blocked by an anticyclone, they may slow to a crawl. Moving over a large land mass, such as Australia, will cause them to weaken considerably within 24 hours due to the lack of water vapour being fed in at the base and the effects of increased friction over the land – which cause the wind to flow more directly into the storm centre, thus increasing the pressure.

Even if they stay over the ocean, their structure will change from that of a tropical cyclone to that of a mid-latitude depression because the cooler sea surface temperatures will diminish the input of heat and moisture. Often they weaken as this happens, but in a small number of cases they can redevelop into a very deep depression capable of heavy rain and storm force winds.

This is often the result when a tropical cyclone moving south meets up with a cold front. As the cold air is drawn into the western side of the tropical cyclone's circulation, the large contrast in density between it and the warm tropical air causes the cold air to slide under and accelerate as it wraps around the centre of the low. This gives the system a new burst of energy. Also, in the upper atmosphere, the trough and jet streams associated with the cold front help intensify the system by taking air away from above the surface depression, thereby lowering its pressure still further.

As the cyclone changes form, it loses its concentric eyewall (although some cumulonimbus clouds are still likely near the centre) and the belt of extreme winds next to the eyewall also weakens. However, the total area of gale and storm force winds actually becomes larger – sometimes by a factor of four. Consequently, although the extreme winds associated with the cyclone weaken, they are replaced by gale to storm force winds often spreading 500 km or more away from the low centre. With a hurricane or tropical cyclone, the damaging winds are usually within about 100 km of the centre and the maximum winds within about 30 km.

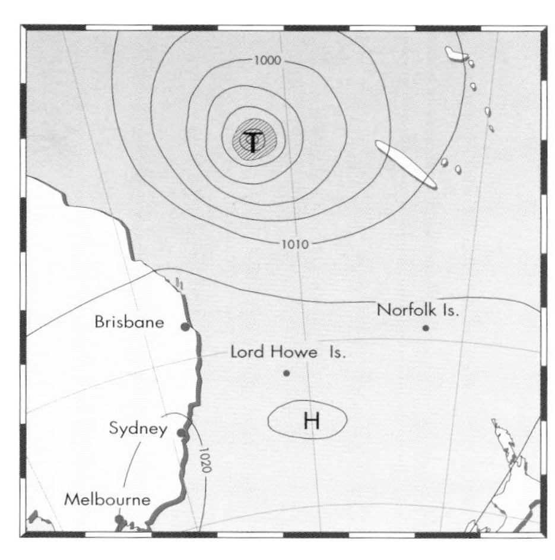

Tropical cyclone Fran, 12 March 1992. The dangerous winds are concentrated in the centre – the shading shows winds over 50 knots. During transition to an ex-tropical cyclone, extreme winds weaken but the area of dangerous winds becomes larger.

This is one reason why these storms are not referred to as hurricanes, or even tropical cyclones, once they reach New Zealand's latitudes. In the tropics they are small enough for ships to avoid if given sufficient warning, but in our part of the world the area of strongest winds is usually too large to be sidestepped and ships must ride them out, unless they can shelter somewhere in the lee of land. The worst time for these ex-tropical cyclones in New Zealand is autumn or the second half of summer. This is because the sea surface temperatures between New Zealand and the tropics remain high into autumn – allowing tropical cyclones to retain their intensity for longer as they travel away from the tropics – and the chance of encountering an outbreak of cold air surging up from near the Antarctic increases as autumn progresses.

Cyclone Bola that flooded Gisborne in March 1988 and cyclone Gisele that sank the *Wahine* in April 1968 were both examples of this process. So too was the great storm of February 1936, which has largely fallen from popular memory, but was arguably the most damaging storm to strike New Zealand this century.

Samoan church destroyed by ocean waves during tropical cyclone Ofa in 1990. Worldwide, 90% of deaths during tropical cyclones are by the sea invading land – known as storm surge.

THE GREAT STORM OF 2 FEBRUARY 1936

This tropical cyclone formed south of the Solomon Islands on 28 January 1936. It met up with a cold front north of New Zealand on 31 January and intensified as it changed to an ex-tropical cyclone, then crossed the North Island on 2 February (fig. 11.5). It was not assigned a name as the practice of naming tropical cyclones did not begin until 1963.

Heavy rain fell over the entire North Island bringing most of the major rivers into flood. One man drowned in Northland when a house was washed away, and another was killed near Thames when his hut was carried into a flooded stream by a slip. Drowned sheep, cattle, pigs and chickens mingled with trees were a commonplace sight in rivers all over the North Island. Roads and railways were cut in many provinces by slips.

An observer by the Ngutunui stream on the side of Mount Pirongia between Kawhia and Te Awamutu saw rimu and kahikatea trees topple end over end in the torrent when their roots or branches caught on the stream bed. When the water subsided, both banks of the stream had been swept clean of soil and vegetation and he picked up 40 dead trout and counted hundreds of dead eels killed by the timber and boulders in the flood.

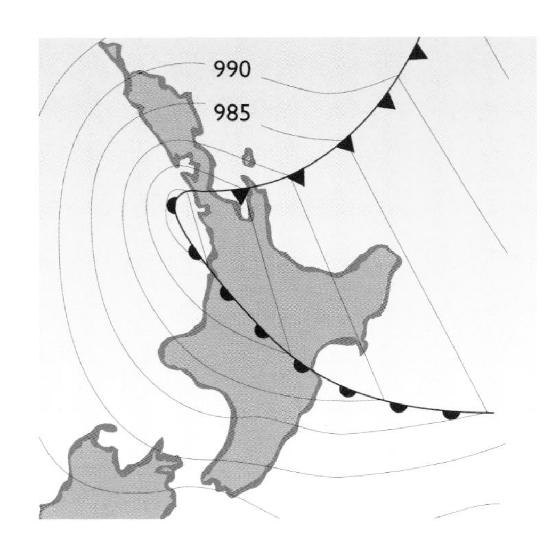

Fig. 11.5. The ex-tropical cyclone that crossed the North Island on 2 and 3 February 1936 was probably the most damaging storm to hit New Zealand this century.

At Kopuaranga in Wairarapa, a 14 ton traction engine disappeared into a river normally only a metre deep and was still missing several days later.

Storm surge caused extreme tides along the east coast of the North Island. Fishing launches were driven ashore at Whitianga in Coromandel. At Te Kaha in the Bay of Plenty, a sea higher than any other in living memory washed a house into the ocean and swept away eight fishing boats. Near East Cape huge seas entered the estuary of the Awatere River and smashed part of a factory, while at Castlepoint on the Wairarapa coast, the sea washed away the sandhills and invaded houses a hundred metres inland.

The wind blew in windows from Picton to Kaitaia and brought down hundreds of thousands of trees, cutting power, telephone, and telegraph lines all over the North Island. Palmerston North was hardest hit. Houses lost roofs, chimneys were blown over, grandstands at three sportsgrounds were demolished (fig. 11.6), and a man was blown off his roof and killed.

A train was derailed near Makerua just south of Palmerston North when three carriages were blown down a bank, injuring a number of passengers. Empty wagons were blown over in several other places and fallen trees blocked the line between Levin and Otaki so that passengers had to take to them with axes before trains could pass.

At Longburn, the Anglican church was demolished and scattered over the road and railway line, while a horse on a nearby farm was cut in half by a flying sheet of corrugated iron.

Fig. 11.6. Grandstand destroyed in Palmerston North

The wind wrought havoc in orchards all over the North Island, destroying much of the fruit. Crops like maize, wheat, and oats were flattened from Marlborough to Northland; haystacks blew away; and in Pukekohe, potato plants were sheared off at ground level. Floodwaters destroyed crops of peas in Marlborough, strawberries and tomatoes in Wanganui, oats in Wairarapa, and kumara in Northland.

In Auckland 40 boats were sunk or driven ashore in the Waitemata Harbour and several more in the Manukau. A fishing launch from New Plymouth was lost at sea, its father and son crew presumed drowned. Numerous small boats were wrecked in Wellington's harbour and a coastal steamer was driven ashore near Kaiwharawhara (fig. 11.7).

Disaster was only narrowly averted when the inter-island ferry *Rangatira* steamed onto rocks on Wellington's south coast in winds almost as bad as those that, 32 years later, would sink the *Wahine*. After being stuck fast for 20 minutes, the *Rangatira* was able to reverse off, then turn and back slowly up the harbour. Taking water in through the gaping holes in her bow, her propellers were half out of the water and her forward lower passenger decks were awash by the time she grounded next to Clyde Quay wharf. Fortunately, none of the 800 passengers and crew suffered serious injury.

Two people died of exposure in the Tararua Range north of Wellington where, at the height of the storm, trees were uprooted from ridges and thrown bodily into valleys. Trampers described whirlwinds twisting the crowns of trees until all the branches splintered off. The dead trunks of some of these can still be seen today.

Among the more unusual effects of the storm was the discovery, at Taupo, of a red-billed tropic bird (amokura) blown down from the Kermadec Islands – which lie about 1000 km northeast of Auckland. Tropic birds are rarely seen in New Zealand, but according to the Victorian ornithologist Buller, Maori in the North Cape area would systematically search the beaches for them after an easterly storm as they valued their red feathers and traded them south for greenstone.

Fig. 11.7. Coastal steamer *John* on rocks at Kaiwharawhara, Wellington.

CYCLONE BOLA – MARCH 1988

Cyclone Bola is mostly remembered for the flooding it brought to the Gisborne district, where it caused an estimated $90 million worth of damage. Rainfall as high as 419 mm in 24 hours was measured at one place and several others had almost 1000 mm over four days. When floodwaters rose and broke through the stopbanks of the Waipaoa River, 3000 people had to be evacuated from the Poverty Bay plains.

Further north near Tolaga Bay, three people drowned after their car was washed off the road while being towed by an energy board truck. But a courageous energy board worker managed to swim across from the truck with a rope tied round his waist and rescue two others. By the time he got them back to the truck, the rising water had entered the cab and it too was in danger of being swept away. They in turn had to be rescued by a school bus driver who managed to get his four wheel drive out to them with the flood waters lapping at the windows.

The railway line to Gisborne was cut when the Waipaoa River washed away a section of the railway bridge (fig. 11.8), and in Wairoa, the bridge carrying the state highway was destroyed. Thousands of sheep were drowned and thousands of acres of crops inundated.

Countless slips and landslides came down and millions of tons of alluvium filled the rivers. When the floodwaters finally receded, silt a metre or more deep covered large areas of farmland and orchards. So much horticultural produce was washed out to sea that fruit was still being dredged from the ocean floor in fishing nets for several months afterwards.

Fig. 11.8. Rail bridge over the Waipaoa River – swept away by floods caused by Cyclone Bola.

A number of factors combined to make the rain so exceptionally heavy. To begin with, the tropical air in the cyclone was very warm and humid and so contained a large

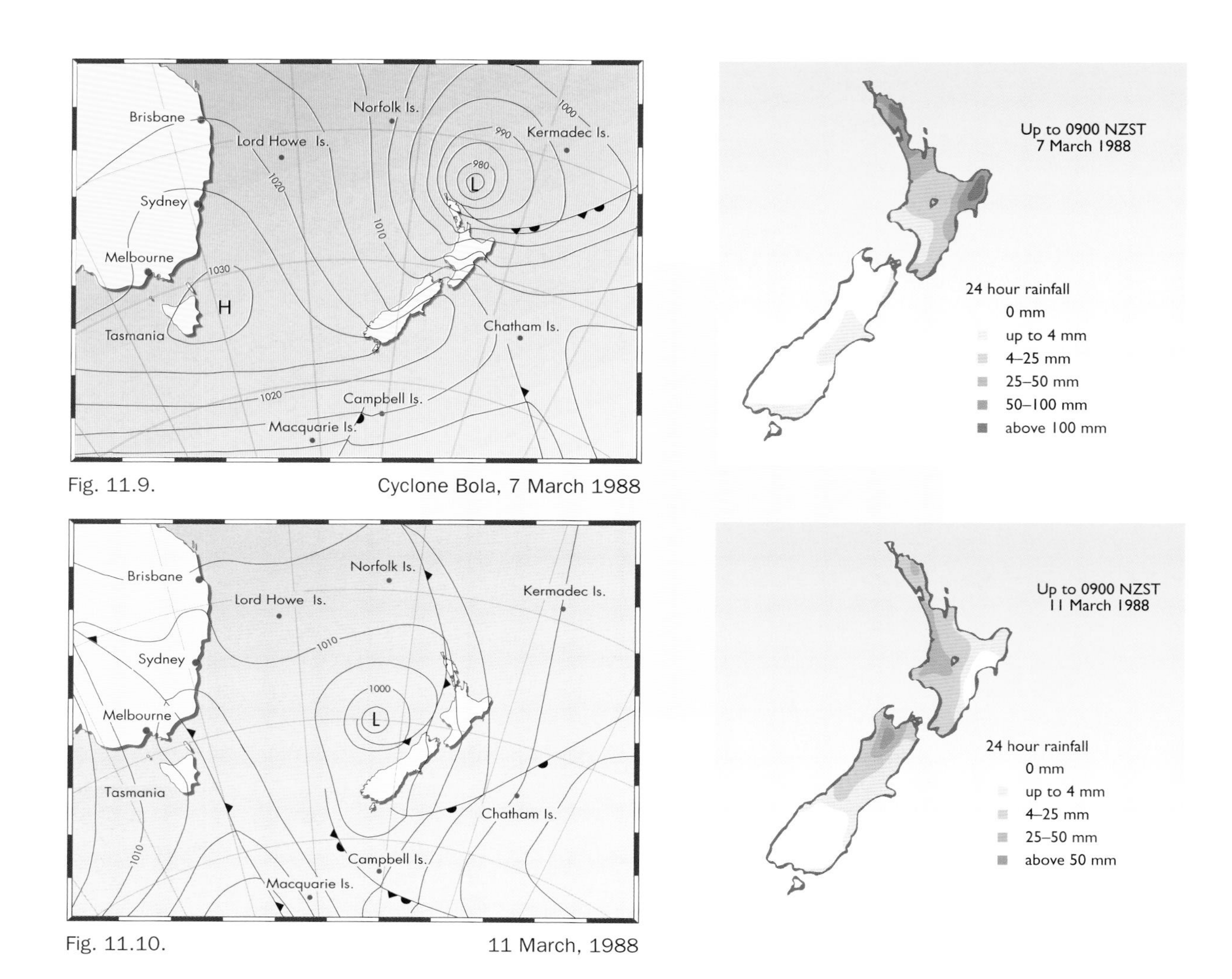

Fig. 11.9. Cyclone Bola, 7 March 1988

Fig. 11.10. 11 March, 1988

amount of water vapour. Then cold air from south of New Zealand happened to be moving up the east coast towards Gisborne as the cyclone approached and, being denser, sank under the tropical air, forcing it to rise (fig. 11.9). The upward air motion was further increased by gale force easterly winds blowing straight into the ranges west of Gisborne. In addition, the cyclone centre moved very slowly when it was north of Gisborne, so the conditions favouring rain persisted for a number of days.

The floods in Gisborne were described as the worst of the century, but it was not the only area that suffered from the heavy rain. Flooding in Northland was said to be the worst in 30 years and thousands of farm animals drowned. It is also interesting to note, however, that some places between Northland and Gisborne had far less rain. Large areas of the Bay of Plenty, for example, had less than a tenth of the rain that fell in Gisborne because they were sheltered by the Raukumara Range. As the air rose over the mountains it dropped most of its rain before or near the main divide, then as it descended on the Bay of Plenty side, it was warmed by compression and dried out further.

Auckland was also partially sheltered by the Coromandel Range and the city received less than half the rain that fell in Whitianga. Similarly, parts of Taranaki and Waikato had almost no rain during the time the wind was blowing from the southeast. But in Northland the wind blew straight off the sea and there was no protection from sheltering hills.

Aside from flood damage, a lot of havoc was also caused by strong winds. These were most violent downwind of the main ranges, so those areas which had been sheltered from the worst of the rain had gales to contend with. The downwind increase occurs because as air flows over a major obstacle like a mountain range, large wave motions are set up in the atmosphere, like those in a stream when water flows over a rock. These waves then bring the stronger winds from higher levels down to the surface.

The pine forests in the central North Island suffered badly with around 17,000 ha of trees uprooted or snapped off. Taranaki was also severely battered with thousands of trees blown over and about 500 houses damaged. In New Plymouth a man trying to repair his roof received multiple fractures when he was blown to the ground and rolled 30 metres down the road. The strong winds were also responsible for a fire getting out of control in the Marlborough Sounds and destroying a large area of native bush.

Areas like Taranaki and Nelson that initially had little or no rain when the wind was blowing from the southeast, were caught when the low centre moved across to the west of the North Island, then south (fig. 11.10). Onshore northerly winds then brought heavy rain, causing several rivers, including the Takaka, to flood.

CYCLONE GISELE – 10 APRIL 1968

The storm that grew out of tropical cyclone Gisele caused the strongest wind ever recorded in New Zealand – 181 km/h (98 knots) gusting 269 km/h (145 knots) at Oteranga Bay west of Wellington. Although it wrought havoc to thousands of properties in Wellington (leading, eventually, to a change in building codes), it is primarily remembered for having sunk the inter-island ferry *Wahine* in Wellington harbour with the loss of 51 lives.

Gisele formed near the Solomon Islands on 6 April 1968, then moved south-southeast towards New Zealand. On the evening of 9 April the low centre was east of Cape Reinga and it crossed the North Island during the night to pass close to Napier at 6.00 a.m. It was about 150 km east of Cook Strait around midday on 10 April (fig. 11.12).

Conditions in Wellington deteriorated rapidly during the morning of the 10 April. As the *Wahine* crossed Cook Strait, the wind rose to 93 km/h (50 knots), the waves increased, and a large drop in pressure was recorded with the barometer falling 5.6 hPa in just over an hour. At this time, some difficulty was experienced in keeping the ship on course, although it was still in deep water and 45 minutes away from the harbour mouth.

Most of the members of the subsequent Court of Inquiry were critical that no consideration was given to abandoning the attempt to enter the harbour at this point. There was a strong likelihood that the waves would be much worse in the shallow waters of the harbour mouth and the steep fall in pressure clearly indicated the possibility of a further increase in the wind strength.

Fig. 11.11. *Wahine*, 10 April 1968.

For reasons never properly understood, though possibly connected with the rain or blowing spray, the *Wahine*'s radar broke down just as she was approaching the harbour mouth. A few minutes later a wave broke over her stern and she sheered off course. After rolling badly off another big wave, the Captain became disoriented in the poor visibility, and manoeuvred the ship in the harbour mouth for almost half an hour before striking Barrett Reef. The ship immediately lost power and both anchors were dropped.

The force of the wind continued to increase and drove the ship up the harbour dragging both anchors. The extreme winds were also forcing extra water into the harbour which, soon after 1.00 p.m. when the wind began to drop, began to flow back out, helping to turn the ship side on to the wind.

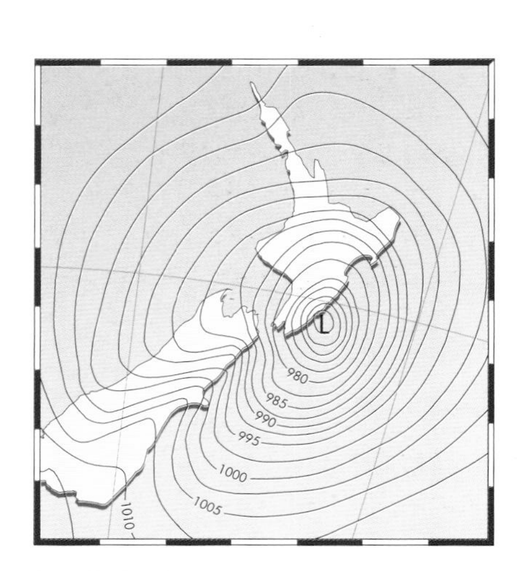

11.12. Tropical cyclone Gisele, 9.00 a.m., 10 April 1968.

This contributed to the list that had developed due to water reaching the car deck where it was able to move freely from side to side. However, it also provided the first opportunity for lifeboats to be lowered on the side relatively sheltered from wind and waves. The order was given to abandon ship. Not long after this was completed, the ship capsized (fig. 11.11).

There is a common misconception that the *Wahine* received inadequate warning about the weather conditions to be expected because of the storm. In fact, this is not true. About the time the ship left Lyttelton on Tuesday night 9 April 1968, she received a forecast for New Zealand coastal waters. The forecast for the central area, which included Cook Strait, was "strong northerlies changing to southerly after midnight tonight, southerlies gradually increasing to gale or storm from tomorrow morning".

Storm force on the Beaufort scale is a mean wind speed of 88–117 km/h (48–63 knots) with gusts 50% stronger, that is gusts 133–176 km/h (72–95 knots). The forecast also included a storm warning describing winds of over 111 km/h (60 knots) around a deep depression of tropical origin that, at that time, was about 96 km east of North Cape and moving south-southeast at 37 km/h (20 knots).

The assessment of the position and movement of the depression were only given as "fair" – meaning accurate to within 1 degree of latitude and longitude. If the forecaster had been more confident about the position and movement he would have described them as "good" – meaning accurate to within half a degree. Had the speed and direction of movement been exact and unchanging, then the area of winds greater than 111 km/h (60 knots) would not have passed over Cook Strait at the time the *Wahine* was due to enter Wellington harbour. In the event, however, the depression did speed up slightly and deepen a little during the night, then change to a more southerly movement about the time the ship hit Barrett Reef.

The fact that the forecast for Cook Strait included the possibility of storm force southerlies of up to 117 km/h (63 knots) gusting 176 km/h (95 knots) was, in part, a measure of the forecaster's uncertainty over the speed and direction of the depression.

At the time the *Wahine* got into difficulty, the wind at Wellington airport had risen to 111 km/h (60 knots) gusting 148 km/h (80 knots) – well within what was predicted. Although the wind continued to rise and peaked at 144 km/h (78 knots) gusting 187 km/h (101 knots) a couple of hours later, the ship was already in serious trouble. Paradoxically, the wind increase may actually have saved lives by blowing the ship back up the harbour: had the *Wahine* rolled over at Barrett Reef rather than within the harbour mouth, the death toll is likely to have been far higher.

CHAPTER 12

What does it all add up to?

Just what to make of New Zealand's weather is often a bit of a puzzle for people from overseas. Dr Fujibe, a Japanese meteorologist who came here in 1997, thought that compared to Japan, New Zealand only had one season: perpetual spring. Juxtaposed against the large swings in temperature that occur between summer and winter in many Northern Hemisphere countries, this is a reasonable comment. Indeed it was controversy over what New Zealand's climate was actually like which gave birth to the first meteorological organisation in the country.

In 1839 a young draughtsman named Charles Heaphy came to New Zealand with the New Zealand Company. After seeing and painting much of the country, he published a short book in London describing his experiences. The book was to some extent a marketing exercise; in it Heaphy claimed that summer in New Zealand lasted five months and winter only two.

He also reassured prospective emigrants that although Maori had acquired firearms, they only used them to shoot in the air at parties. Ironically, Heaphy was to win the Victoria Cross 25 years later for bravery under fire in the Waikato Land Wars. But his generous description of New Zealand's climate was soon also under fire as disillusioned settlers who had read his propaganda wrote home complaining of the incessant wind and changeable weather.

Famous for his exploration of the South Island's West Coast (above), as well as his landscape painting, Charles Heaphy was one of the first to publicise an opinion of New Zealand's climate.

This led to, in 1856, the colonial administration collecting and publishing the first weather statistics for New Zealand to prove that the climate was not as grim as some settlers were claiming.

Observatories were set up soon after, and in 1861 the Auditor General, Dr Charles Knight, was made the first Director of Meteorological Services.

Certainly, New Zealand's lack of a hard European winter made a strong impression on most immigrants and was trumpeted far and wide by enthusiasts such as William Pember Reeves, New Zealand's high commissioner in London, who attributed the success of the 1905 All Blacks to the beneficial effects of growing up in New Zealand's healthy climate.

NEW ZEALAND'S HEALTHY CLIMATE

Climate is the sum total of a multitude of individual weather events, and is often described in terms of averages – such as the average rainfall for the year or the month. Knowledge of such threshold levels is invaluable for engineering purposes. Knowing the

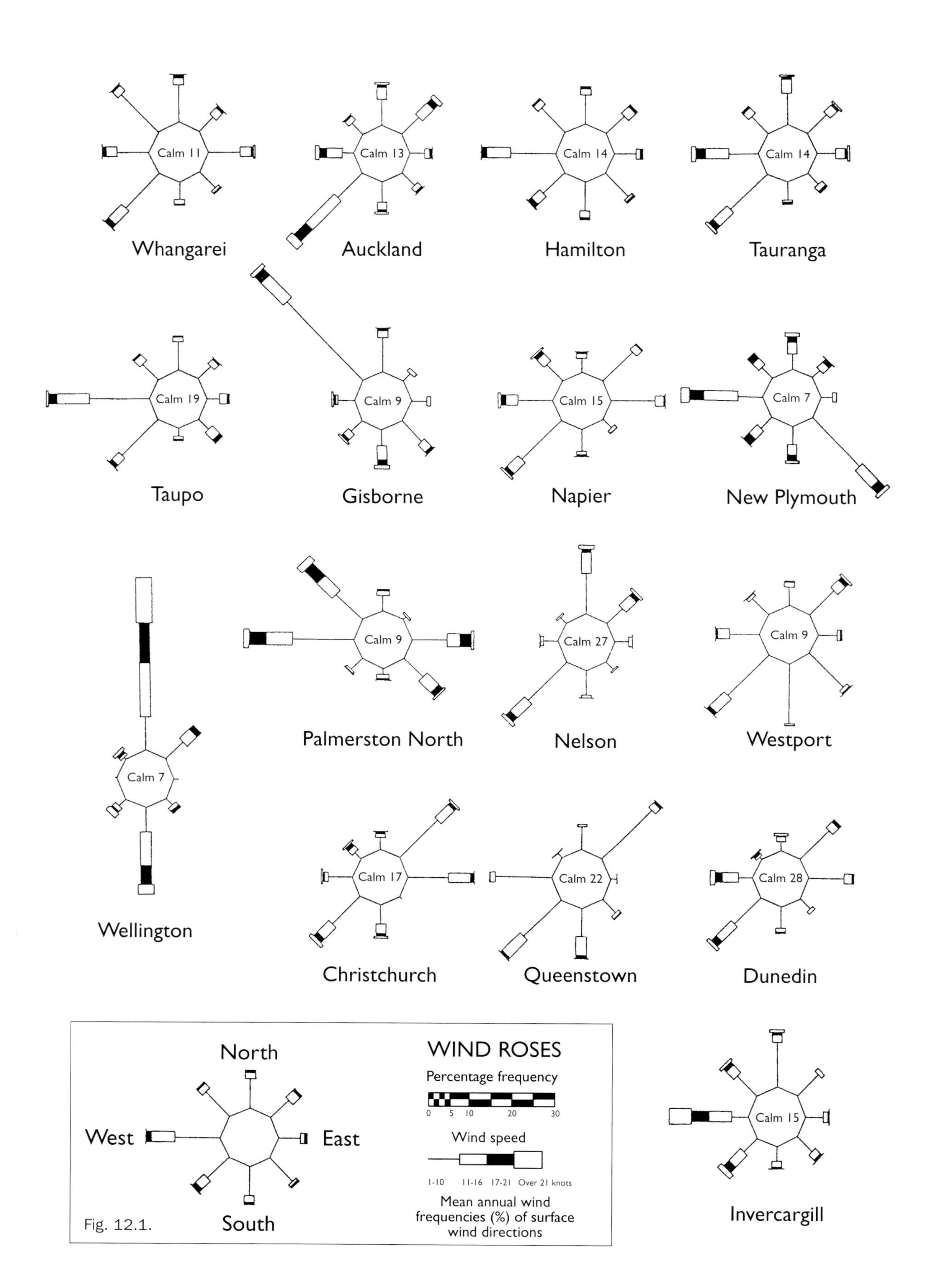

Calm 11
Whangarei
Calm 13
Auckland
Calm 14
Hamilton
Calm 14
Tauranga
Calm 19
Taupo
Calm 9
Gisborne
Calm 15
Napier
Calm 7
New Plymouth
Calm 7
Wellington
Calm 9
Palmerston North
Calm 27
Nelson
Calm 9
Westport
Calm 17
Christchurch
Calm 22
Queenstown
Calm 28
Dunedin
Calm 15
Invercargill
North
West
East
South
WIND ROSES
Percentage frequency
0 5 10 20 30
Wind speed
1-10 11-16 17-21 Over 21 knots
Mean annual wind frequencies (%) of surface wind directions

Fig. 12.1.

heaviest 24 hour rainfall that an area could expect in, say, a 10, 50 or 100 year period is vital information when a structure such as a bridge is being planned.

Wind A useful way to depict the frequency, strength and direction of wind in a particular location over a period of time is a wind rose (fig. 12.1). The length of each arm is proportional to how often the wind blows from that direction, while the length of the different segments show how often the wind is of a particular strength.

As New Zealand lies astride the belt of westerly winds that predominate in the mid-latitudes around the world, most of the wind roses in figure 12.1 show winds prevailing from the northwest, west, or southwest.

The wind roses also reveal the effects of the local terrain; most dramatic is the case of Wellington where the westerly wind is blocked by the wall of the Tararua Range and deflected through Cook Strait as a northerly. Equally, all the winds that try to approach Wellington from a southwest to easterly direction tend to be channelled into southerlies by the Inland and Seaward Kaikoura ranges.

The effect of the land can also be seen in the Gisborne wind rose where the prevailing wind is a northwesterly. The reason is the broad Waipaoa Valley, which extends away to the northwest of Gisborne City and channels most types of westerly airstreams down its length. In addition, on calm, clear nights a katabatic wind develops and flows down the valley and over the city.

New Plymouth experiences a frequent katabatic wind from the southeast, which also tends to be the wind direction whenever a good southerly is blowing through Cook Strait. Both of these effects are due to Mount Taranaki and together they make southeast the most common wind direction in New Plymouth.

Sea breezes make significant contributions to the wind in coastal locations, as can be seen in the northerlies at Nelson, and the northeasterlies at Napier and Christchurch. The northeasterly is the most frequent wind in Christchurch and is aided by the presence of a lee trough, such as the one shown in figure 9.4. This lee trough northeast wind also affects the rest of the South Island east coast.

The wind roses also clearly show how the wind tends to be stronger in coastal locations like Auckland, Wellington and Invercargill, while inland centres such as Hamilton and Queenstown have the lightest winds. This is because moving air encounters more friction over the land than the sea. Palmerston North with its strong westerlies is an exception to this. The reason is the combined effects of the Southern Alps and the Manawatu Gorge. If the air is stable and unable to rise over the ranges, it spills around the northern end of the Southern Alps, across Wanganui and Palmerston North and through the Manawatu Gorge. Palmerston North also experiences frequent easterly winds funnelled through the Gorge from the other side.

Rainfall The map of the average annual rainfall (fig. 12.2) clearly shows the effect of the mountains on the westerly winds which prevail in New Zealand's latitudes. The highest totals occur west of the Southern Alps, with parts of Fiordland experiencing over 7000 mm per year, while across the mountains in Central Otago some places get less than 400 mm annually.

The pattern over the North Island is a little different, reflecting the fact that most of the rain in the westerlies actually comes from the northwest. Consequently, northern Taranaki, which is exposed to the northwest, averages 1600 mm of rain a year, but most of the Manawatu, which is sheltered from the northwest, has less than 1200 mm. There is also an area of maximum rainfall stretching across the high ground of the central North

Island from Mount Taranaki to the Raukumara Range and East Cape, while the lowest values occur near Napier, which is in the lee of these ranges during a northwest flow.

Another difference between western and eastern areas is that the *variability* of rainfall from year to year is greater in eastern areas in the lee of the ranges. That is to say eastern areas experience bigger swings between wet and dry years, whereas western areas tend more consistently towards their average annual rainfall.

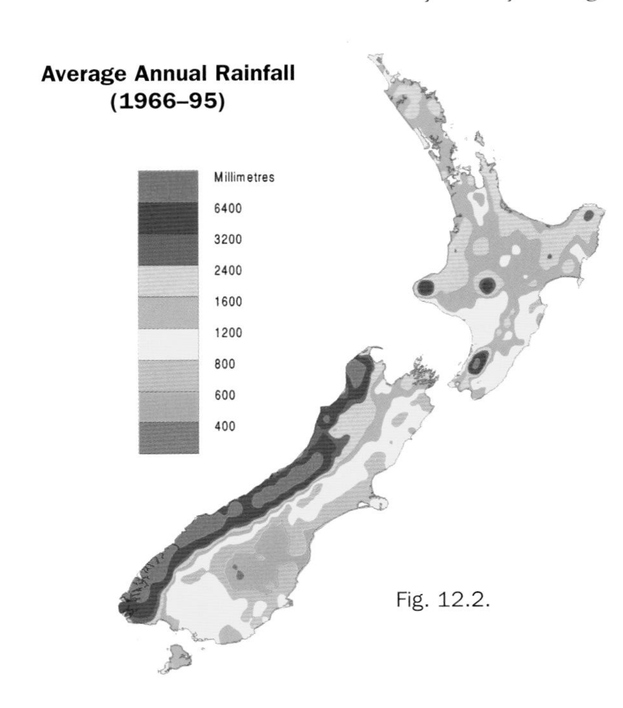

Fig. 12.2.

Another way of looking at rainfall is to consider the average number of days a year that more than one millimetre of rain falls (fig. 12.3). The pattern is similar to the average rainfall map except that the maximum rainfall across the North Island high ground is not so clear cut. This is because during northwest rain events, rain still tends to fall in areas either side of the high ground even though it will be much heavier in the mountains.

Temperature Although air temperature varies markedly between day and night as well as from day to day, a map of average annual temperature is still helpful in portraying New Zealand's weather. Figure 12.4 clearly shows higher temperatures in the north of the North Island where the sunlight is more intense due to its increasingly direct angle of entry. Also, of course, it is colder in the mountains and warmer lower down because air temperature decreases with height through the atmosphere. As we saw in Chapter 1, the atmosphere receives most of its heat from the Earth's surface, but as air rises it expands and cools.

The daily range of temperature varies from place to place, being generally greater east of the main ranges where westerlies often produce clear skies, allowing a frosty night to follow a warm day. In fact the highest temperatures in New Zealand have been recorded on the east coast of the South Island in föhn nor'westers, even though the average temperature for these areas is lower than most of the North Island.

MOUNTAIN CLIMATE

As a consequence of the air being colder the higher you go in the mountains, it is also more likely to be cloudy and raining. As a rule of thumb, when the weather is deteriorating the rain usually begins in the mountains, where it will be heavier and last for longer in comparison to surrounding coastal areas. In fact, there are many occasions when it will only rain in the mountains, such as when there is sea breeze convergence in the Tasman mountains of northwest Nelson or in the Raukumara Range between Gisborne and Bay of Plenty.

Conversely, there are also occasions when an anticyclone may bring cloud and drizzle to low levels, while above the inversion layer (created by the descending air in the anticyclone) the mountains are clear as a bell.

Wind strength is another aspect of weather that tends to be more extreme at high levels – to the extent of easily being capable of blowing a person or, on rarer occasions, a hut off a mountainside. Extreme winds are likely when the weather map is covered in isobars (e.g. figs. 2.1 and 11.5), but can also occur near anticyclones when the weather map looks rather more innocent. For example, the situation on 25 March 1992 (fig. 9.2) produced storm force westerlies of 93 km/h (50 knots) about the tops of the Ruahine

Days of Rainfall

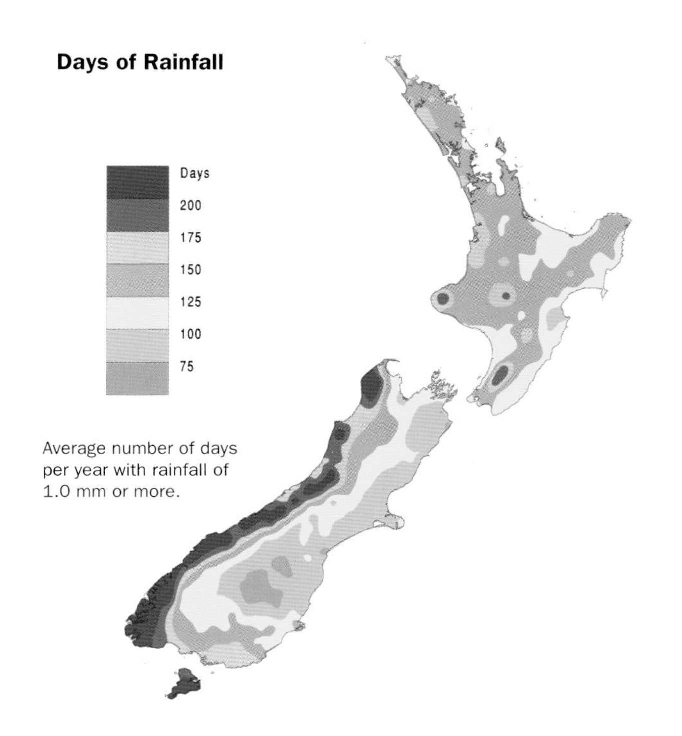

Fig. 12.3.

Mean annual temperature, 1966-95

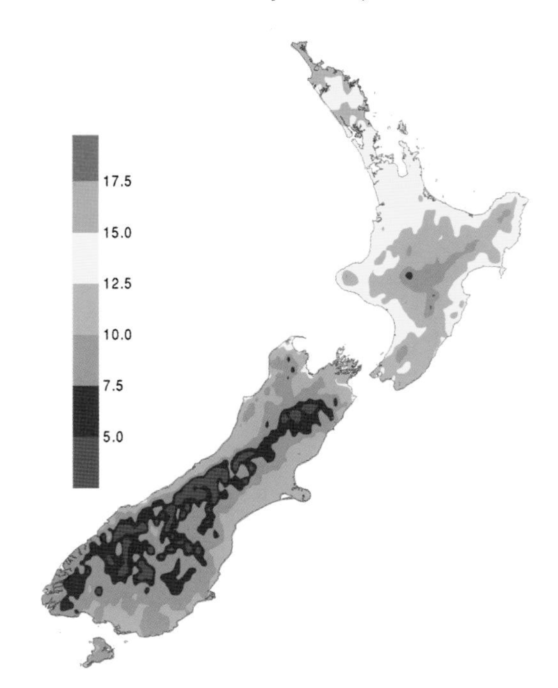

Fig. 12.4.

Range while the wind strength in nearby lowland areas was less than half this.

The example is a typical illustration: the atmosphere was stable because of the anticyclone and there was an inversion layer just above mountain top level. The air rising over the mountain chain thus became squeezed between the stable air of the inversion and the tops of the mountains and accelerated, a bit like how water flowing gently from a tap can be squirted across a room if you hold your thumb over the mouth of the tap.

Frequently, air rising rapidly over a ridge will cool enough to form cloud, reducing visibility to a matter of metres and making it easy to become lost. The combination of wind and low temperatures also increases the risk of exposure when travelling along the tops.

Aside from the dangers of exposure and losing one's way, drowning in flooded rivers is the other main weather hazard in the mountains. Rivers can rise surprisingly quickly and it's important to appreciate how light rain lower down could be torrential a few kilometres away near the headwaters of the river.

It is also worth noting that a semi-swollen river can be more treacherous than one in full flood because no one would sensibly step into a river in full flood. It is the river that could perhaps be crossed successfully nine times out of ten that most deceives.

EL NIÑO AND THE SOUTHERN OSCILLATION (ENSO)

The year to year fluctuations in the various weather parameters such as rainfall and temperature are of great interest, particularly to those involved in agriculture. Once a closed book, some recognisable patterns are now emerging, allowing a certain amount of seasonal forecasting to be attempted. The most important of these patterns is the Southern Oscillation, sometimes known as El Niño, and referred to by meteorologists as ENSO (El Niño-Southern Oscillation). ENSO is a weather pattern centred in the ocean and atmosphere of the tropical Pacific, but affecting a large part of the world. In New Zealand it contributed to the major droughts of 1982–83 and 1997–98, as well as to the Auckland water crisis of 1994.

Fig. 12.5. This map shows that during the 1997–98 El Niño, sea surface temperatures over the eastern tropical Pacific were 3–4 degrees warmer than normal.

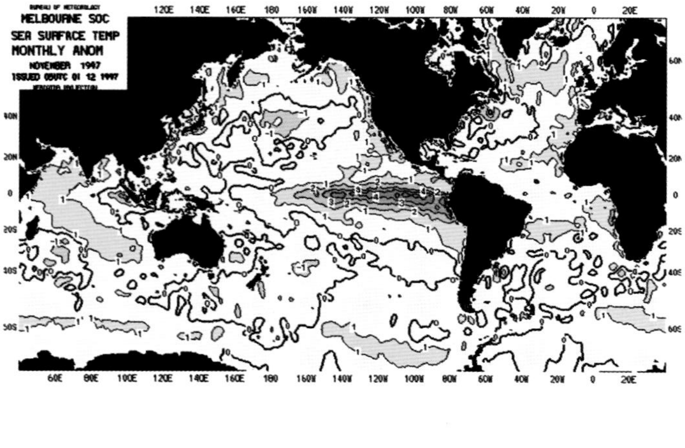

In the 1920s, an English meteorologist named Gilbert Walker noticed that when the monthly average air pressure between Tahiti and South America was lower than average, there was a strong tendency for higher than average pressures around the New Guinea/Darwin area. Conversely, when pressures around Darwin were lower than average, Tahiti's pressure tended to be higher than average.

This seesawing of pressure between the east and west of the Pacific is known as the Southern Oscillation. When Darwin's pressure is up and Tahiti's is down, the Southern Oscillation is said to be in its negative phase, known as El Niño. When Darwin's pressure is down and Tahiti's is up, the Southern Oscillation is said to be in a positive phase, known as La Niña.

ENSO and Ocean Temperature The ocean also plays a crucial role in the Southern Oscillation. During an El Niño, the sea surface temperatures in the tropics are higher than average by 2°C or more from around the dateline all the way to America, and colder than average on the Australian/Indonesian side of the Pacific (fig. 12.5). During a La Niña, the sea surface temperatures are above average on the Australian/Indonesian side of the Pacific, and below average on the South American side.

When the Southern Oscillation is weak and there is neither an El Niño nor a La Niña, sea surface temperatures are warmer on the western side of the tropical Pacific and colder on the South American side. This is caused by the easterly winds (known as the trade winds because their consistency of direction was a boon to trading ships in the days of sail) that blow in the Tropics for most of the time, piling up warm water on the Australian side of the Pacific. The trade winds actually increase the sea level by about 40 cm on the Australian side compared to the American side.

Because the sea surface is warmer in the west, a lot of upward air motion occurs as heat transferred from the ocean to the air causes the air to expand and become less dense and more buoyant. While the air rises, it cools by expansion, causing water vapour to condense into cloud and rain. But as it condenses, the water vapour releases the heat previously transferred to it by the warm ocean into the air: therefore the air does not cool as fast as it might otherwise have, and so remains buoyant and keeps rising.

Once the rising air reaches the top of the atmosphere, it spreads out; some moving away from the tropics to mid-latitudes like New Zealand, and some moving across the tropical Pacific and descending over the colder ocean waters near South America. The areas of desert along the coast of South America are a direct result of this downward air motion suppressing cloud and rain.

During an El Niño, the situation is partially reversed. The sea surface temperatures rise in the eastern Pacific, and the upward air motion on the Australian side weakens and becomes intermittent so that rainfall there is much diminished. Meanwhile, upward air

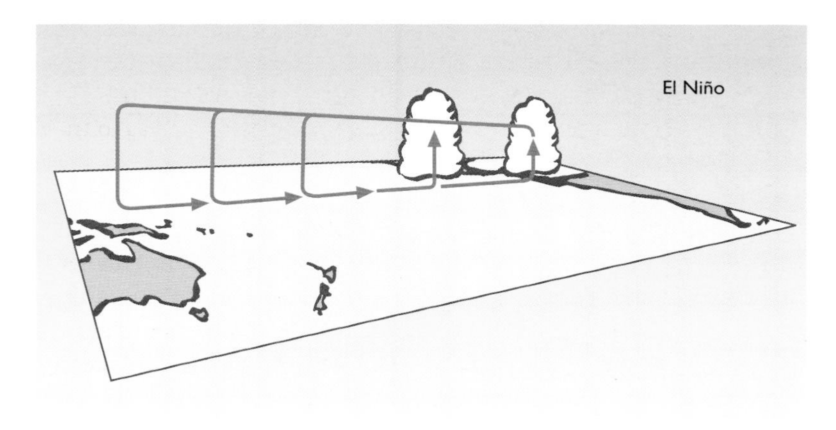

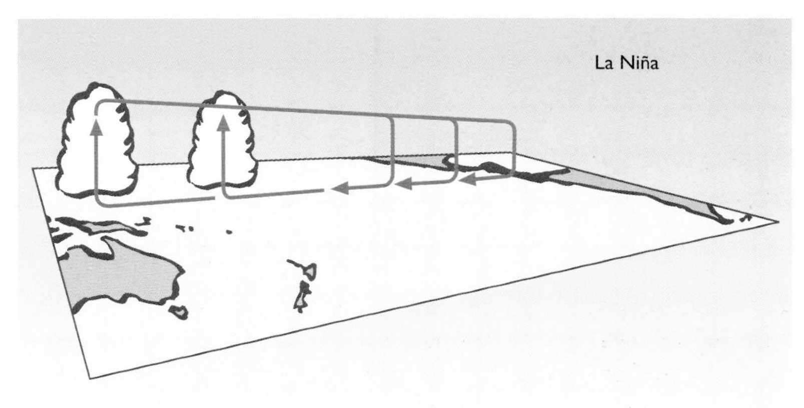

Fig. 12.6. Upward motion in the tropics tends to occur over the warmest part of the ocean. During an El Niño this is shifted over the eastern Pacific, while during a La Niña it remains in the west but is stronger than normal.

motion, and therefore rainfall, shifts eastwards across the tropical Pacific, sometimes as far as the Americas (figs. 12.6 and 12.7).

During an El Niño, the increased upward air motion over the American side of the Pacific plus the diminished and intermittent upward air motion over the Australian/Indonesian area, act to decrease – or occasionally reverse – the easterly trade winds. And this in turn helps to maintain, or even strengthen, the warm anomaly in sea surface temperatures on the American side of the Pacific.

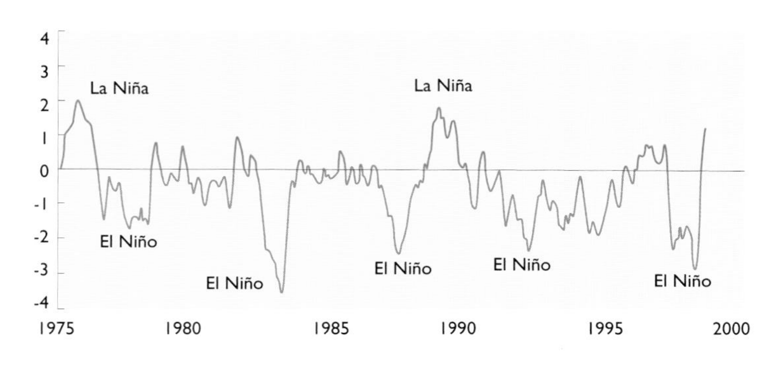

Fig. 12.7. Southern Oscillation Index. The index is based on the pressure difference between Tahiti and Darwin and shows the strength of La Niña and El Niño events. The last 20 years have been dominated by frequent strong El Niños, but some climatologists expect this to change in the near future.

Similarly, during a La Niña the increased upward air motion in the west and downward air motion in the east causes the easterly trade winds to strengthen. And this provides a feedback mechanism that helps to maintain, or even strengthen, the sea surface temperature difference across the Pacific.

It is this linking of air and ocean anomalies through the action of the trade winds that allows the Southern Oscillation to stick in either a La Niña or El Niño pattern for many months. When eventually one phase ends, the system tends to swing towards the other extreme, and then back again, in an irregular oscillation taking between three and six years to go from one El Niño to the next.

El Niño Weather When it comes to forecasting the weather, the useful thing about the Southern Oscillation is that if it is strongly positive or negative in spring, then it tends to stay that way throughout the summer.

A strong negative Southern Oscillation (El Niño) correlates with more frequent southwesterly flows over New Zealand and a ridge of high pressure to the north. The summer of 1982–83 was the strongest El Niño event of the last hundred years – as measured by the difference in air pressures between Darwin and Tahiti – while the El Niño of 1997–98 was the strongest, as measured by the change in sea surface temperatures. Weather situations like the one for 5 February 1983 (fig. 12.8) were much more frequent during both summers, with the stronger than normal westerly winds leading to cooler conditions, especially in western areas.

Rainfall for both summers was well above average in Southland, Westland, Fiordland

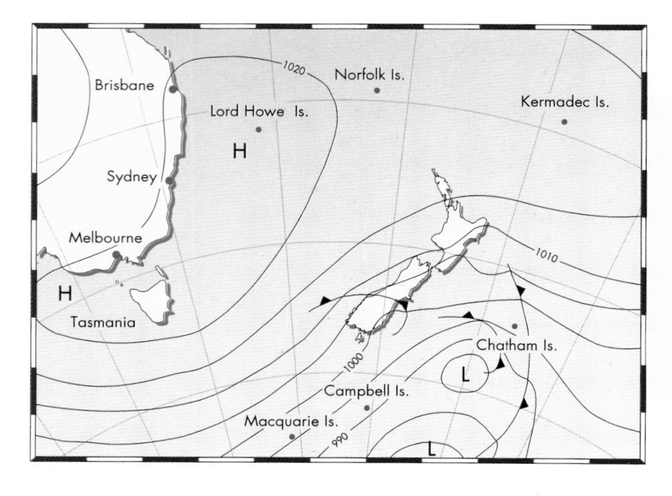

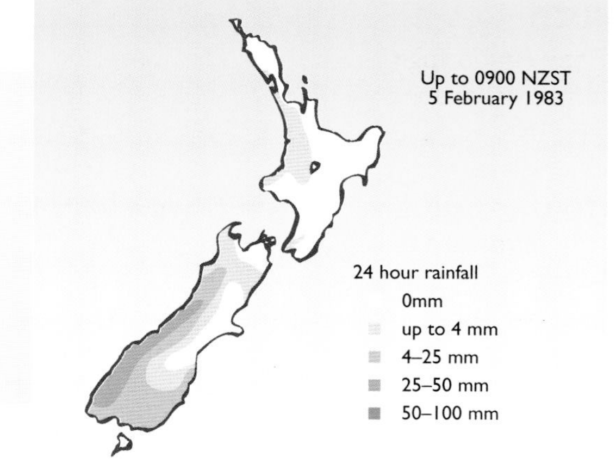

Fig. 12.8. 5 February 1983. Typical El Niño weather map and rainfall distribution.

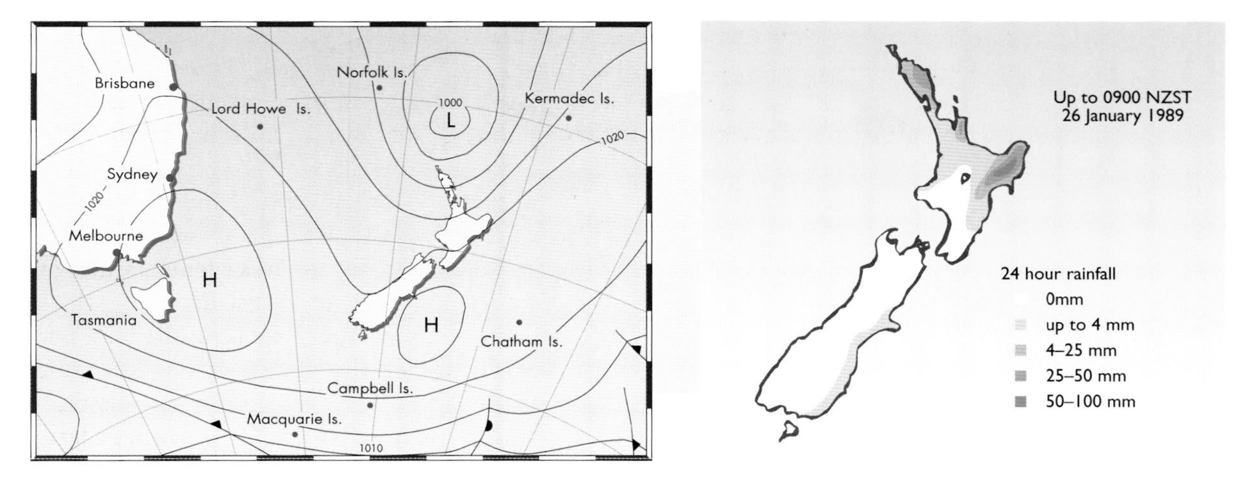

Fig. 12.9. Typical La Niña weather map and rainfall distribution, 26 January 1989.

and parts of Otago, where fronts often passed over when the wind was blowing from sea to land thus enhancing rainfall. Over the rest of New Zealand rainfall was below average; particularly in Northland and Auckland which were under the influence of a ridge of high pressure, and in eastern districts from Gisborne to Canterbury which were in the rain shadow of the western ranges.

But although the Southern Oscillation exerts a strong influence on New Zealand's weather, it does not dictate everything that happens so much as indicate a trend. You cannot blame the fall of every raindrop, or their failure to appear, on El Niño and La Niña. Some studies indicate that it explains only about 20% of the year to year variations in rainfall. So, for example, during the 1982–83 El Niño there was severe drought in Gisborne and Hawkes Bay, but during the 1987–88 El Niño there was not.

The Southern Oscillation affects the entire Pacific region and India as well as, to a lesser extent, areas beyond such as Africa and North America. During a strong negative Southern Oscillation (El Niño), the monsoon rains over Indonesia, northern Australia and sometimes India, are much reduced, causing drought, food shortages and forest fires. On the eastern side of the Pacific where there are large areas of desert, rainfalls tend to be increased, often resulting in disastrous flooding.

La Niña Weather During a strong positive Southern Oscillation (La Niña), the warmer than average temperatures on the western side of the tropical Pacific intensify the upward air motion, bringing more rain than normal. In Bangladesh and Australia this has resulted in disastrous flooding. The strong upward air motion delivers more air to the top of the atmosphere, increasing the outflow across the Pacific and thereby also the downward air motion on the American side. This suppresses the few cloud systems that might have brought rain to the already arid regions there, increasing the likelihood of drought.

A strong La Niña correlates with more frequent northeasterly weather over the North Island and anticyclones in South Island latitudes. The summer of 1988–89 was a good example of a La Niña event, and situations like the one shown for 26 January 1989 (fig.12.9) were more common than normal. On the map, a depression can be seen near the top of the North Island and the northeast winds associated with it are bringing warm, humid air down over the North Island resulting in heavy rain over Northland, Coromandel and the Gisborne ranges. Little or no rain fell from Taranaki to Wairarapa southwards, either because of sheltering or their close proximity to the high.

Over the whole summer, many areas from Gisborne northwards had roughly twice their normal amounts of rain, while rainfall further south was less than average. The frequent northeasterlies also meant that temperatures were above average over most of the country.

ENSO and Global Climate Because an El Niño is associated with warmer sea surface temperatures over a large area of the tropics, the average global temperature rises. In New Zealand, however, the temperature falls because cold, southwesterly winds are more common.

Conversely, during a La Niña a large area of the tropical ocean is cooler than average and therefore the average global temperature drops about half a degree. Over New Zealand, however, temperatures are warmer than average because warm, northeast airstreams occur more often than normal.

Recently, the Southern Oscillation was stuck in an El Niño phase that lasted almost six years from 1990 to 1995. This is the longest El Niño on record, prompting fears that it could be a sign of climate change brought about by the increasing levels of carbon dioxide and other greenhouse gases in the atmosphere. However, climatologists have also detected a long term swing in ENSO activity that suggests El Niños should become weaker and less frequent sometime in the next decade

CLIMATE CHANGE: THE GREENHOUSE EFFECT

One of the major scientific and political issues of our times is the possibility that human activities are changing the composition of the atmosphere in ways that will lead to a significant shift in the Earth's climate towards warmer temperatures. The atmosphere's natural greenhouse effect already keeps the Earth's surface temperature some 30°C warmer than it would otherwise be, making life possible.

The greenhouse effect works in the following way. Much of the sun's radiation is in wavelengths in the visible part of the spectrum; these are able to pass relatively unhindered through the atmosphere to warm the Earth's surface. The warm Earth then radiates energy outwards, but because the Earth's temperature is much less than that of the sun, the characteristic wavelength of its radiation is much longer.

Much of this outgoing long wave radiation gets captured by molecules in the atmosphere and a significant amount is re-radiated back down to further warm the Earth's surface. Therefore any increase in the concentrations of the gases that can capture and re-radiate long wave radiation will enhance the greenhouse effect.

Water vapour is the most important greenhouse gas, followed by carbon dioxide (other contributing gases include methane, nitrous oxide and chlorofluorocarbons). Carbon dioxide occurs naturally in the atmosphere and plays an important part in the lives of most organisms. Animals and humans breathe it out, while plants take it in during photosynthesis to form the energy sources which sustain them.

By measuring the amount of carbon dioxide trapped in air bubbles in various layers of the Greenland ice sheet, scientists have been able to show that the concentration of carbon dioxide in the atmosphere has been steadily increasing over the years (fig. 12.10). Before the Industrial Revolution it is estimated that carbon dioxide in the atmosphere amounted to 280 parts per million. Current measurements show it to have risen by more than 25%.

The burning of fossil fuels (gas, coal, oil) is the major cause, but destruction of the world's native forests has also contributed. The widespread burning off of trees in the preparation of ground for agriculture, as well as releasing more carbon dioxide into the atmosphere, has also reduced one of its primary 'sinks' because trees absorb carbon dioxide in order to grow. Another important sink for carbon dioxide is the ocean where, for example, marine blooms of phytoplankton absorb large amounts of carbon dioxide.

Current estimates indicate that future levels of carbon dioxide could reach double the pre-industrial level by the year 2050 if concerted attempts are not made to cut emissions.

Estimating the effects of increased amounts of greenhouse gases on climate is complicated by a number of feedback mechanisms. For example, the warmer it gets, the more water will evaporate into the atmosphere. Water vapour is an important greenhouse gas, so increasing the amount in the atmosphere will lead to further warming. This amplification of the greenhouse effect is yet another positive feedback mechanism in the climate system.

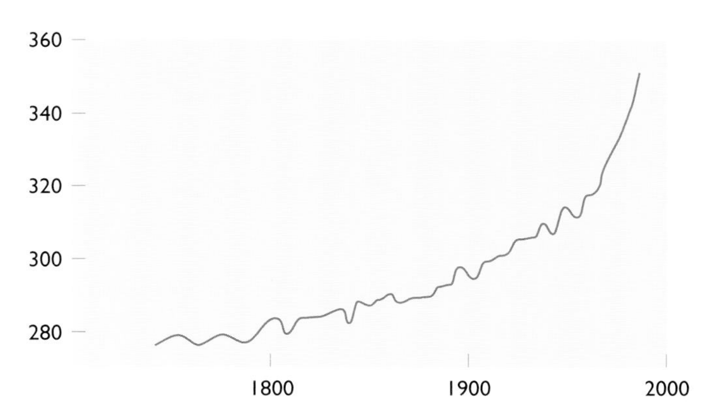

Fig. 12.10. Graph showing the steady increase in amount of carbon dioxide in the atmosphere since the industrial revolution.

On the other hand, the greater the concentration of water vapour, the greater the likelihood of clouds, and these will reflect some of the incoming sunlight straight back into space, helping to cool the Earth. This is a negative feedback mechanism.

As temperatures rise, part of the Arctic tundra permafrost will thaw releasing methane, a potent greenhouse gas, which will cause further warming. This is another positive feedback mechanism.

Taking all this into consideration, the Intergovernmental Panel on Climate Change (IPCC – a joint body formed by United Nations and the World Meteorological Organisation) has predicted that, if nothing is done to limit the emission of greenhouse gases, then the average global temperature will rise by 1°C by the year 2025 and by 3°C by the year 2100. This is a faster rise in average global temperature than has occurred at any stage during the last 10,000 years.

As result of the warming, mean sea level is predicted to rise by 65 cm by the end of the next century – mainly due to thermal expansion of the surface waters as they warm, plus the melting of some land ice.

What will the consequences of these changes be? There is no better way to begin to answer this than to look back at past changes and see how these have affected human history. But first of all, how do we know what the past climate was like?

How is climate change measured? The methods used to study past climates are as varied as they are ingenious. One of the most important involves the ratio of two different isotopes of oxygen known as oxygen 18 and oxygen 16. Oxygen 16 is normally about 500 times more abundant, but the crucial difference is that oxygen 18 is heavier (because it has two more neutrons in the nucleus) and is consequently unable to evaporate as readily as oxygen 16.

Accordingly, the proportion of oxygen 18 in the water vapour in the atmosphere is slightly less than the proportion in sea water. The extent of this effect, however, depends on the temperature of the sea when the water evaporates. Therefore, from the ratio of oxygen 18 to oxygen 16 found in the annual layers of snow laid down on the ice caps of Greenland and Antarctica, a measure of the temperature of the tropical oceans at the time the water evaporated can be worked out.

The same approach can be used with oxygen extracted from the carbonates from skeletons of surface-dwelling plankton found in sea-bed cores. The oxygen isotope ratio indicates the temperature of the water the organism lived in, then a variety of dating techniques can be used to establish the age of the skeleton as far back as a million years.

On land, fossil pollen is one of the most important indicators of past climate as it is extremely long lasting and readily identified. Pollen preserved in sediments thousands of years old show which plants were growing at that time. Plants have different sensitivities to rainfall, temperature and frost, so a record of what was growing enables a reconstruction of past climate.

Tree rings are another important source of information about rainfall and temperature. Roughly speaking, trees growing on the edge of an arid region will have a large growth ring in a year with above average rainfall, and trees growing near the edge of their temperature tolerance, say halfway up a mountain, will have a large growth ring in a warmer than average year.

Tree rings can be dated to an actual year by counting back from the present. This is easy if the tree is still alive and if not, scientists can overlap trees of different ages to go back thousands of years. The oldest living tree, a bristlecone pine in the White Mountains of California, is about 4300 years old!

In some parts of the world, lake sediments are mostly laid down in spring when rivers are at their highest. These sediments form distinct layers called varves, the thickness and composition of which can then be used to estimate the spring rainfall. One lake in Sweden has layers that can be counted back 15,000 years to around the finish of the last ice age.

The nature of the sea-floor sediment also tells a tale: fine dust is windblown but certain small rock fragments can only have reached present locations by being carried and dropped by melting icebergs.

Historical anecdotes are also important in fleshing out a past climate picture. For example, at the height of the Little Ice Age around AD 1700, Eskimos in kayaks were sighted off Scotland on several different occasions. And in the *Landnámabók*, written in Iceland in AD 1125, there is the story of Thorkel Farsek, one of the first Greenland settlers. Around AD 990 he is reputed to have swum more than 3 km across a fiord to bring home a sheep to feed a relative who visited unexpectedly. It is estimated that the lowest temperature he could have survived would have been 10°C. Today, the sea temperature there in summer is about 6°C.

Around New Zealand the colder temperatures of last century made it possible for icebergs to reach the Chatham Islands. Several grounded there in 1892 and stayed long enough for part of one to be chipped off and made into a pot of tea by the locals, before it drifted away again with a change of wind.

What has the climate been doing? The Earth's climate in the last 2.5 million years has been dominated by ice ages separated by brief warm periods known as interglacials (fig. 12.11). Ice age conditions have prevailed for about 90% of the time and there seem to have been 20 or so warm interglacial periods, such as the one we live in now. What started the cycle of ice ages 2.5 million years ago remains a matter for investigation and speculation, but one plausible theory is that they followed as a consequence of changes to the ocean currents when South America joined up with North America at Panama and the waters of the tropical Atlantic were prevented from flowing into the tropical Pacific.

The swings into and out of ice age conditions occurring roughly every 120,000 years seem to correlate with variations in the Earth's orbit around the sun known as the

Milankovitch cycles. The Earth's orbit is not circular but elliptical, and the ellipticity varies and repeats every 100,000 years. The tilt of the Earth's axis of rotation relative to the plane of its orbit is what gives us the change of seasons from summer to winter. This tilt varies from 22 degrees to 25 degrees in a cycle taking 40,000 years. The greater the tilt the greater the temperature difference between summer and winter.

As the Earth rotates about its axis, it is also slowly wobbling in a similar way to a spinning top that is tilted a little to one side. This is known as precessing and has the effect of varying the seasonal position of the Earth on its way around the ellipse over a period of 20,000 years. At present the Earth is closest to the Sun in the Southern Hemisphere summer, therefore the intensity of the sun's radiation is greater in a Southern Hemisphere summer than a Northern Hemisphere summer. (Which is one reason why it is easier to get sunburnt in New Zealand than at comparable latitudes in the Northern Hemisphere.)

The changes in the intensity of the sun's radiation at the Earth's surface caused by orbital variations are alone not enough to explain the temperature differences between an ice age and an interglacial. A feedback mechanism is needed to amplify the effect – this is provided by the extended cover of snow and ice that exists during an ice age, from which most of the sunlight reaching the Earth is reflected, thus depriving the Earth of a significant amount of energy.

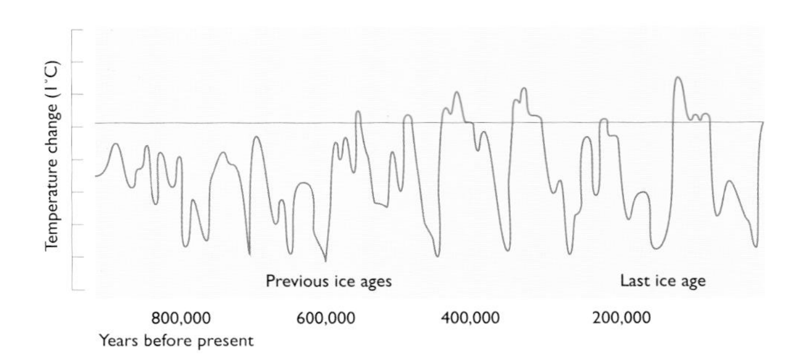

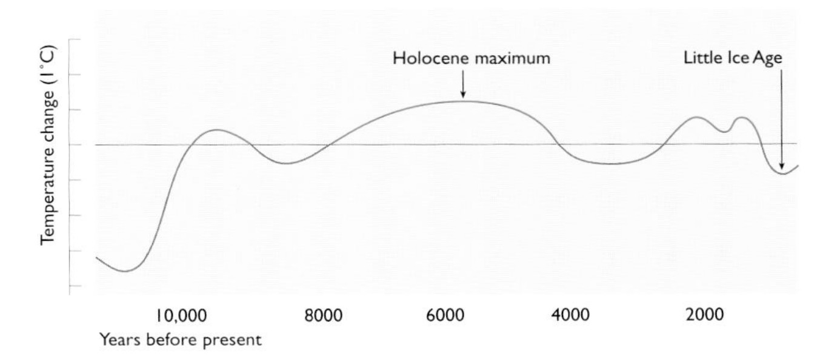

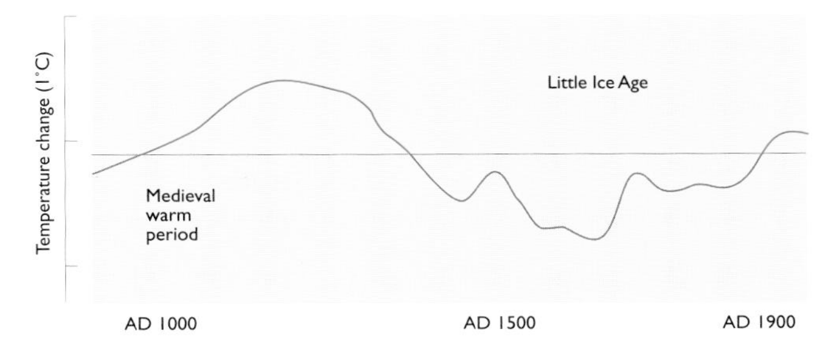

Fig. 12.11.
Measurements of past climate show a great deal of natural variability, not only over the 100,000 year timescale of major glacial events, but also over the shorter term. These have all had dramatic influences on human history. The central line through each graph represents conditions at the beginning of the twentieth century.

The key factor in triggering the change from an interglacial to an ice age is, therefore, the extent to which the winter snow cover persists through summer. If a cold, cloudy summer follows a cold winter then a significant amount of snow may last through to the next winter. The more intense the summer sunlight, the harder it is for the snow cover to last. A situation where blocking anticyclones were positioned to bring cold, cloudy weather to snow covered continents would aid the persistence of the snow cover.

Another, more random, cooling influence on climate are volcanic eruptions. An eruption which is big enough and directed upwards (as opposed to, say, Mount St Helens which largely went sideways) will lift enough material into the stratosphere to affect global climate. Although much of the ash and dust falls out of the atmosphere within months, sulphate aerosols can stay in the stratosphere for years. If the eruption occurs between latitude 20 degrees north and 20 degrees south, the aerosols can spread over both hemispheres as they migrate slowly poleward. Aerosols block some of the sun's incoming radiation but do not trap the Earth's outgoing long wave radiation.

The largest volcanic eruption in the last 200,000 years was the massive Toba eruption in Sumatra 80,000 years ago. Recent calculations suggest that this may have caused land temperatures in a zone 30 degrees north to 70 degrees north to be 5 to 15°C colder than normal and dramatically intensified the onset of the last ice age.

The last ice age began about 80,000 years ago and lasted until about 10,000 years ago. At its peak, 18,000 years ago, the average global temperature was about 5°C colder than at present, with the average temperature in New Zealand about 8°C colder.

In the Northern Hemisphere an extensive ice cap covered Scandinavia, Scotland, most of England and much of Northern Europe and Canada. So much water was locked up in the ice that sea levels were 100 m lower than at present. Thus England was joined to Europe, and Asia to North America. In New Zealand a land bridge connected Nelson to Taranaki and over much of the country the climate was similar to that prevailing over the subantarctic islands today.

Fig. 12.12. In Wairarapa a 45,000 year old kahikatea stump emerges from beneath 40 m of shingle.

The weather during the ice ages played a major role in shaping New Zealand as we now know it. Erosion in the mountains was more pronounced because the tree lines were lower and frost action more vigorous in breaking up rocks. Consequently, the rivers brought vast amounts of shingle down from the mountains and left deposits hundreds of metres thick, forming areas like the Canterbury and Wairarapa plains (fig. 12.12).

Strong winds caused dust storms that transported millions of tons of river silt over many parts of New Zealand in characteristic rolling dune shapes. In the downlands of South Canterbury and North Otago these deposits – known as loess – are as much as 10 m deep (fig. 12.13).

As the ice sheets melted at the end of the last ice age, the sea level rose as fast as five metres a century. Rather than a gradual encroachment, it is more likely that an inundation of coastal plains would have occurred in a series of catastrophic spring tide storm surges, such as those documented in Europe during the late Middle Ages. These events may have given rise to the 'great flood' legends common throughout the world.

Fig. 12.13. Deep layers of windblown loess deposited during the ice ages revealed at Dashing Rocks, Timaru.

Once the ice sheets had melted, parts of the Earth's surface formerly depressed by the great weight of ice began to rise. Scandinavia, for example, has risen by about 300 m and is still rising today, as is Scotland. This rise is contributing to sinking in the south of England as the plate tilts. Such rebounding of the Earth's surface makes diagnosing past sea levels a lot trickier.

During the last ice age, the lakes and inland seas of temperate and lower latitudes were far larger and deeper than today. Lake Chad, on the southern boundary of the Sahara desert, was 40 m deeper and as large as the United Kingdom. In Utah, North America, cars contest the land speed record on the Bonneville Salt Flats where there was once a lake 300 m deep covering 50,000 square km.

Today the Sahara Desert is one of the driest places on Earth, yet it contains hundreds of rock paintings that show its arid landscape was once rich in plants and animals. Completed between 5000 and 15,000 years ago, these paintings depict elephants, rhinoceros, hippopotamus, crocodiles, and even fishermen in boats. Nor are they the only sign of a wetter climate: radar scans from space have penetrated the dry sands to reveal the remnants of ancient river systems.

After about 3000 BC, huge areas of Asia, Africa and Arabia started to dry out and people began migrating to the river valleys of the Nile, the Euphrates and Tigris, the

Indus and the Huang Ho. It seems possible that the biblical account of the wanderings of Abraham could be a record of these migrations.

The drying continued into the second millennium BC and civilisations like Harappa and Mohenjo-daro that had developed in the Indus river valley (in present day Pakistan) were brought to an end by drought around 1700 BC.

About 1626 BC, a catastrophic volcanic eruption occurred on the island of Santorini in the Aegean Sea. The amount of rock and ash thrown into the atmosphere is estimated to have been five or more times greater than that of the Krakatoa eruption in AD 1883. Average global temperatures may have been lowered by as much as 1–2°C for several years and the effect in high latitudes is likely to have been even greater.

Pollen analysis shows that the forest in Canada retreated rapidly around this time, finishing 200 to 400 km from the northern limit achieved in 4000 BC. The withdrawal was accompanied by widespread fires, presumably started by lightning. Within a century the whole zone was covered in tundra.

There is abundant evidence to show global cooling continued into the last millennium before Christ. Glaciers advanced in many mountainous areas; tree lines retreated downhill. Trade routes over the Austrian Alps appear to have been closed and high altitude gold mining ceased.

The effects of the cooling can be seen in early agricultural records. For instance, a comparison between the records of old Babylon from 1800 to 1650 BC, with those of later Babylonia from 600 to 400 BC, found that the average date for the beginning of the barley harvest was more than a month later.

Glaciers retreated when the Earth's climate warmed at the end of the last ice age, although this general retreat has been punctuated by temporary advances, such as the one that formed the Waiho Loop (a terminal moraine) 11,000 to 12,000 years ago on the plains below Fox Glacier.

At the beginning of Roman times temperatures began to rise again. The Roman historian, Livy, records that around 300 BC, the river Tiber froze over several times and that beech trees were growing near Rome. In the first century AD, the writings of Pliny tell us that the Tiber no longer froze, the beech was regarded as a mountain tree and that vines and olives were being grown farther north in Italy than ever before. Repeated invasions of Europe by nomadic peoples from the east such as the Huns, the Goths and the Vandals, marked the decline and fall of the Roman Empire. The invasions seem to have been driven by the drying up of pastures over the steppes between Russia and China. The Caspian Sea dropped to very low levels at this time, and the Silk Route to China fell into disuse as settlements along its length were abandoned due to drought.

Temperatures cooled in the centuries after the Romans, but warmed again towards the end of the first millennium AD. In the warmth of the tenth century AD, the Vikings crossed the North Atlantic, which was relatively free of ice, and established settlements in Iceland, Greenland and North America. The North American settlement was short-lived, partly due to the hostile reception from the Indians, but also because of climate change.

The onset of a colder period known as the Little Ice Age in the thirteenth century gradually rendered the sea routes impassable with extensive sea ice. The settlement in Greenland eventually died out and the population of Iceland was greatly reduced. It is interesting to speculate that had this cooling not occurred, the Vikings might have returned to North America and the people of what is now the United States might be speaking Norwegian today.

In Europe the onset of the colder weather was abrupt. A string of very wet, cool years with poor harvests occurred from 1313 to 1321. In the summer of 1315 harvests failed to ripen at all, resulting in famine with large scale loss of life. There were even reports of cannibalism. Combined with the Black Death later in the century, this pattern of weather led to the eventual abandonment of tens of thousands of villages on higher ground across Europe.

The bubonic plague epidemic, known as the Black Death, may itself have had its origin in a weather event. In 1332 exceptional rainfall in China brought flooding that killed an estimated 7 million people. In the aftermath there would have been an explosion in the rat population. The plague, which was endemic in China, was carried by a flea that lived on rats. It quickly spread across the Asian continent into Europe. In the space of 18 months, one third of the world's population from India to Iceland had died.

Although the sea level was a little lower in the fourteenth and fifteenth centuries than it is today, the increased storminess of the Little Ice Age caused a number of catastrophic storm surge floods along the west coast of Europe with death tolls of 100,000 or more. Some of these formed the Zuyder Zee in Holland, which was not drained until this century. Parts of England suffered coastal erosion and the two ports of Ravenspur and Dunwich disappeared into the sea.

Among the events marking the peak of the Little Ice Age was the 30 year failure of the cod fishery around the Faroe Islands to the west of Norway. Although cod like cold water, their livers cannot function below 2°C, so the absence of cod around the Faroes shows that sea temperatures there must have been about 5°C colder than they are today.

Colder sea temperatures in the sixteenth and seventeenth centuries also caused the herring fishery to shift away from Norway into the North Sea near Holland and England – to the economic benefit of those nations. Although temperatures began to recover during the eighteenth century, there were still a number of bad years. Significantly, a number of poor harvests preceding the French Revolution in 1792 caused the price of bread to rise, so that by the eve of the revolution, Paris workers were spending 88% of their income on bread alone.

A series of sizeable volcanic eruptions took place near the end of the eighteenth century and, in 1815, an enormous eruption occurred at Tambora in Indonesia. This brought "the year without a summer" in 1816 when snow fell in June over a wide area of eastern North America. Harvests were dramatically reduced over much of Europe and northern India, exacerbating the typhus epidemic of 1816–19, which was the most extensive in European history. Famine also opened the door to the plague, which raged through southeast Europe and the eastern Mediterranean at that time, and the first great epidemic of cholera, which started in Bengal in 1816–17 and swept the world. The cold, stormy weather in Switzerland that summer also encouraged Mary Shelley to write *Frankenstein*.

The Irish famine in 1846 was fuelled by weather conditions favourable to the spread of potato blight. The fungus responsible for the blight multiplies rapidly during periods of successive days with temperatures of 10°C or greater and relative humidity of 90%. That summer, Ireland had a predominance of warm, moist, southerly winds. Over six years the famine, combined with a typhus epidemic, killed around a million people and triggered a wave of migration to the United States.

Impacts of Climate Change Before looking into New Zealand's future climate, it is useful to consider some of the effects that weather and climate can have on plants and animals.

Booms and crashes in the insect and animal populations have been recorded as far back as biblical times, and it is likely that the weather had a role to play in triggering many of them. One of the ways this happens is through the way the weather affects the plants that provide food for animals.

For example, the flowering and seed production of New Zealand's beech trees varies considerably from year to year (fig. 12.14). In years of great abundance – known as 'mast years' – there can be a fifty-fold increase in production of seeds. In mast years this 5000% increase in seed production may affect all three species of beech at the same time, along with other plants. Although the trigger mechanism is not completely understood, there is strong evidence that hot, dry conditions in late summer and autumn will lead to a mast year in the following spring and summer.

The abundant food supplies in mast years provide excellent conditions for the breeding of forest birds such as kaka. Indeed, in years when seed production is poor, these birds do not even attempt to breed.

A good example of how the weather can influence birds' breeding occurred in the Galapagos Islands near Ecuador during the 1982–83 El Niño. Persistent torrential rains caused an explosion of plants and seeds so that young female finches began breeding at three months instead of the usual age of two years.

Fig. 12.14. Flowering and seed production of New Zealand beech forests is influenced by weather and climate.

Mast years in the South Island beech forests are also good years for mice. The trees' abundant flowering in spring is followed by a rapid increase in the number of caterpillars. Some of these fall to the ground and become food for the mice, who then breed more successfully. When the beech seeds fall a couple of months later, the mice benefit again. Eventually, their population becomes so large that their food resources become exhausted and the mice begin to migrate. When this happens, roads become dotted with flattened mice and trout have been found gorged with those that have either drowned or been eaten alive as they were attempting to cross streams.

A recent example of rodent-borne disease occurred in the United States when dozens of people living near the Four Corners region were killed by an outbreak of hantavirus. Again the weather was involved as heavy rains and snows in the spring of 1992 in New Mexico caused the desert to bloom, leading to an abundance of piñon nuts and grasshoppers. This was followed by a tenfold increase in the population of deer mice, who were the carriers of the infection.

Food supplies are not the only way the weather can affect animals. After cyclone Andrew devastated parts of Florida and Louisiana in 1992, an estimated 182 million fish died of suffocation. The winds had thrown so many trees and branches into the inland waterways that oxygen levels became depleted as the organic matter decayed, causing the fish to suffocate.

But the interaction between weather and animals is often more subtle than this. For example, it has long been known that browsing animals, such as giraffes, travel slowly upwind as they feed. It had been thought that this was to move away from potential predators downwind. However, it has now been discovered that certain trees, when eaten, rapidly alter the chemicals in their leaves to make them unpalatable and even toxic to animals. Some of these chemicals escape into the air as the leaves are crushed and other trees downwind are able to sense them and begin altering their own chemistry before they too are attacked.

NEW ZEALAND'S FUTURE CLIMATE

So what might the future hold for New Zealand's climate? Unfortunately, going from a prediction of an increase in average global temperature of 1°C by 2025, to predicting the change in temperature over New Zealand is not so straightforward. This is because disruptions to the global wind patterns are also likely, which could result in some regions experiencing the opposite of the global temperature trend. It is too early to say exactly which regions these will be, but the way in which an El Niño produces colder than average conditions over New Zealand but warmer than average global temperatures, is an example of the problem's difficulty.

Although the changes predicted in mean temperature are small compared to the range of temperatures experienced in any one day in New Zealand, the possible consequences for plant life are considerable. The southern limit for growing particular crops and trees should shift by about 200 km. However, crops which require winter frosts to promote flowering and fruiting, such as apricots, peaches and nectarines, would no longer be profitable in northern areas that became frost-free.

Growing kiwifruit could become uneconomic in the Bay of Plenty, but could extend as far south as Canterbury. Then again, warmer conditions in the north could allow the farming of new crops such as bananas, pineapples, or lychees.

Lack of frost and warmer, more humid conditions may see an increase in outbreaks of stock diseases such as salmonella, facial eczema, and footrot, while large lamb losses in the snow and rain of a springtime cold snap could be rarer. Weeds, including varieties such as kikuyu grass that are already present in New Zealand, may spread explosively.

Although it is possible that warmer temperatures may bring a net benefit to New Zealand agriculture, the pace of change is likely to be too fast for native vegetation and some species may become extinct. Confined to pockets of land surrounded by farms, the seeds of native species may be physically unable to reach new climatic niches – which are likely to be hundreds of kilometres away or no longer even exist. Species such as kauri, which prefer dry summers, could well be reduced to tiny pockets of habitat, while quick growing, adaptable introduced species, such as old man's beard or *Pinus contorta*, may take over large areas.

CHAPTER 13

Forecasting

HISTORY

As far back as you can trace history you find humans grappling with the weather in efforts to use, anticipate, or influence it. Lines scratched on pieces of animal bone over 30,000 years ago have been interpreted as moon counts made by humans to measure the passage of time, and hence the seasons. For it is the progression of the seasons, and the weather that comes with them, that controls the flow of food; be it animal migrations, the ripening of crops, or the fruiting of trees.

The most powerful of the ancient gods were often those who could control the weather. Sacrifices would be made to persuade them to withhold damaging storms or bring lifegiving rain for crops, and favourable weather for harvests, hunting expeditions and times of war.

Nowhere more than in the history of warfare, does such a clear picture emerge of people taking advantage of the weather and forecasts to decisively influence events. One of the major turning points in the history of the ancient world was the defeat by the Greeks of the invading Persian forces of Xerxes at the naval battle of Salamis in 479 BC. Themistocles, commander of the Athenian fleet, used local knowledge of the intricacies of the sea breeze in the strait between the island of Salamis and the mainland to help outmanoeuvre the Persian fleet and win a crushing victory.

At the battle of Towton in 1461, during the Wars of the Roses, Edward IV's army attacked in a snowstorm with the wind at their backs. His archers used the wind to carry their arrows further into the ranks of the Lancastrian army and took advantage of the blowing snow to shield their movements. The enemy archers could not clearly see that Edward's archers were advancing a dozen yards to fire then retreating after each volley, so the return fire was falling short.

Ten years later at the battle of Barnet, early morning mist helped Edward and his men to victory when the right wing of the Lancastrian army wrapped around them, only to be decimated by its own archers who were unable to distinguish friend from foe in the mist.

After Edward died, his brother Richard III seized the throne and sought to discredit Edward. Among his many accusations was one that Edward had used the well-known necromancer, Friar Bungay, not only to conjure up the mist that had helped him win the battle of Barnet, but also to unleash a number of storms in the English Channel that had shipwrecked various of his enemies.

In New Zealand, weather and forecasting have also played a crucial role in warfare. During the Land War in south Taranaki in 1868, the Ngaruahine leader, Titokowaru, had some 80 warriors at his disposal, compared to nearly 1000 in the colonial army under Lieutenant Colonel Thomas McDonnell. In order to provoke McDonnell into pursuing him into the bush, Titokowaru attacked a redoubt at Turuturu Mokai near Hawera, only 5 km from the main camp of the colonial army. He chose to attack just before dawn on a

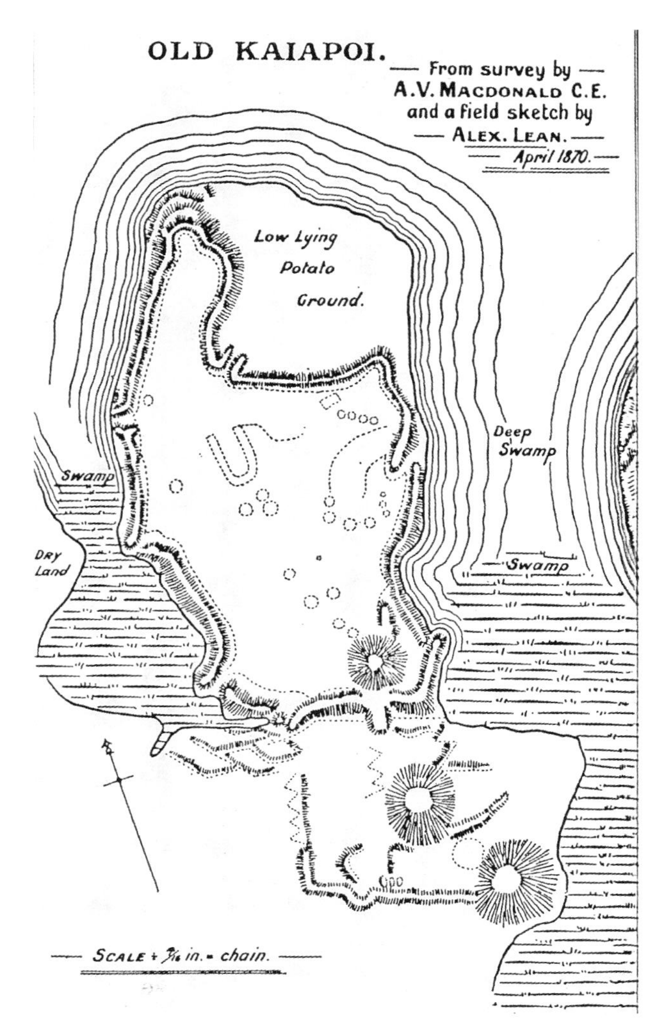

Fig. 13.1. Kaiapohia Pa.

day when he correctly predicted a strong westerly wind, so that the sound of gunfire would not reach the main camp – a mere 20 minutes cavalry ride away. The attack was successful, with 16 colonial casualties as against six for Ngaruahine. Two months later, McDonnell attacked Titokowaru's stronghold in the bush and suffered one of the greatest defeats in New Zealand history.

The weather also played a decisive role when Te Rauparaha besieged Ngai Tahu in their pa at Kaiapohia (old Kaiapoi) in 1831. Built on a peninsula in a swamp, the pa could be attacked from only one side, which was protected by a strong palisade (fig. 13.1). Te Rauparaha's men dug deep zigzag trenches up to the palisade and, under cover of darkness, stacked manuka high against it so as to burn their way in. But before setting it alight, they needed a southerly wind to blow the flames against the palisade.

Instead, a northwest wind developed. The Ngai Tahu defenders knew that a northwest wind in this area was almost always followed by a southerly, so they gambled and set fire to the manuka themselves, hoping to burn it out before the southerly arrived. But the wind change came quickly, the palisade burned, and the pa was taken with considerable loss of life.

FITZROY

One of the founders of modern weather forecasting, Robert FitzRoy, also played a crucial role in Te Rauparaha's life. Appointed Governor of New Zealand, he arrived in New Zealand just after the Wairau Incident in June 1843 and had to adjudicate the affair. He judged that Te Rauparaha was wrong to have executed prisoners after the fighting, but that the settlers had no right to attempt to arrest Te Rauparaha and certainly not with force. For this judgement he was burned in effigy on the streets of Nelson.

FitzRoy is a good example of an applied scientist of the Victorian era. In his youth, he captained the *Beagle* on a voyage around South America, on which Charles Darwin made the discoveries that led to his theory of evolution. He was one of the first to fit lightning conductors to his ship and to carry a barometer to measure atmospheric pressure in order to forecast the weather.

On returning to England after his recall as Governor of New Zealand, FitzRoy carried two barometers with him. When the ship anchored for the night in the Strait of Magellan the weather was fine and calm but both FitzRoy's barometers were falling fast and he believed a storm was imminent. FitzRoy tried to persuade the captain to take extra precautions but he refused and retired to his cabin. FitzRoy then turned his attention to the first mate and eventually, late at night, persuaded him to put out a second anchor on a heavy chain. Several hours later a storm struck the ship with such force that it snapped the heavy anchor chain, which then whiplashed away and became tangled in the lighter chain. It was this tangle, however, that held the ship, saving her and all on board from being dashed upon the rocks.

Back in England, FitzRoy returned to the navy and rose to the rank of Rear Admiral. In 1857 he was appointed Chief Statist to The Board of Trade and commissioned to research and compile weather statistics.

In those days thousands of lives were lost each year in shipwrecks. For example, in the great storm of 1709 – as documented by Daniel Defoe – an estimated 8000 lives were lost at sea! (This storm also incinerated many of the windmills in England by spinning their sails so fast that their wooden gears overheated from the friction and burst into flames!)

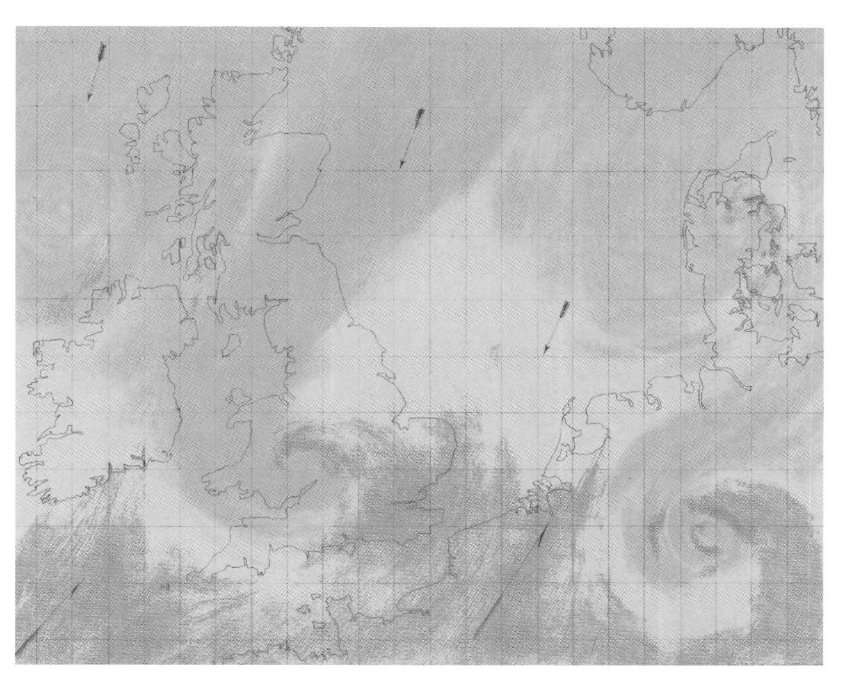

Fig. 13.2. Drawn in the 1860s, FitzRoy's illustration of the interaction of polar and tropical air masses looks remarkably like a modern satellite photo. This map appeared in FitzRoy's *The Weather Book – a manual of practical meteorology*, published in 1863.

From his own success in using a barometer, FitzRoy believed passionately in the possibility of developing scientific methods of forecasting that could save lives. The invention of the telegraph made the rapid exchange of weather observations from over a large area possible, and then, based on the swift analysis of these observations, the rapid dissemination of forecasts.

After a while, FitzRoy began to issue weather forecasts to the daily papers in London as well as to the House Guards and the Royal Humane Society. He also wrote a book describing his methods – in particular the idea that storms in the westerly winds were brought about by the convergence of tropical and polar air masses moving away from their source regions into the mid-latitudes (fig. 13.2). This idea foreshadowed one of the major advances of twentieth century meteorology when Norwegian forecasters came up with the idea of cold fronts and warm fronts concentrating bands of cloud and rain at the boundaries between cold and warm air masses.

Although FitzRoy's forecasts were well received by the fishing community, he was eventually criticised in Parliament for their accuracy and for exceeding his brief. Also at this time, FitzRoy, a devout Christian, was upset by the emergence of Darwin's theory of evolution – which he felt partly responsible for, having been the one who gave Darwin the chance to take part in the voyage of the *Beagle*. At one of the passionate and sometimes acrimonious public debates about evolution, FitzRoy was heard to mutter that mammoths had only become extinct because the doors of the Ark were too small!

Tragically, all of this pressure contributed to a nervous breakdown, and FitzRoy's suicide.

THE FORECAST FOR D-DAY

There is perhaps no greater vindication of FitzRoy's belief in the worth of weather forecasting than the accurate forecast for the invasion of Europe by the Allies on 6 June 1944. Good weather for the landings was crucial to the invasion, which, had it failed, could not have been attempted again for another year. Had the war been prolonged, the increased death toll could well have been measured in millions.

Rain fell from overcast skies and gale force winds drove large waves onto the beaches of Normandy as dawn broke on Monday 5 June 1944. To the Germans watching from their defences, there was nothing to show that this was the day the Allied Armies had planned to invade Europe. In fact, the operation had been put on hold because the bad weather had been forecast 24 hours before. Had it gone ahead in these conditions, the invasion would have been a catastrophic failure.

Nevertheless, the invasion had to occur on either the 5th, 6th, or 7th of June to take advantage of the right conditions of moon and tide. Darkness was needed when the airborne troops went in, but moonlight once they were on the ground. And a spring low tide was necessary so that the landing craft could spot and avoid the thousands of mined obstacles on the beaches. If this narrow time slot was missed, the invasion would have to be delayed for two weeks.

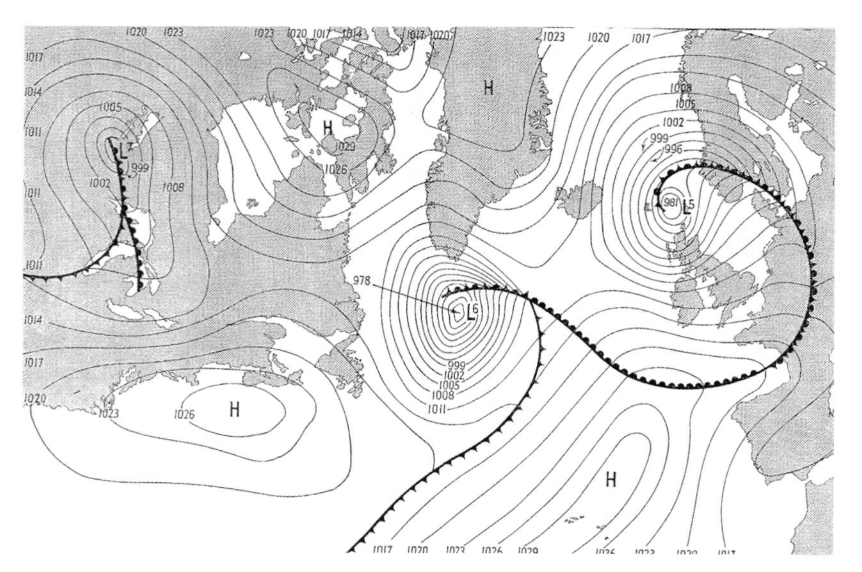

Fig. 13.3. The weather on June 5 1945 – the day originally planned as D-Day – would have been catastrophic for the invading force. The brief period of acceptable weather came with the ridge intensifying behind the front over France.

The decision to postpone the invasion for 24 hours had been taken by Eisenhower and the Supreme Command at 0430 on Sunday 4 June. It was not taken lightly; so many ships were already at sea converging on the Normandy coast that the risk of detection was grave.

Nor had the forecast which prompted the postponement been easily arrived at. Eisenhower's weather advice was provided by Group Captain Stagg, a forecaster seconded from the British Meteorological Office, who was in turn coordinating the advice of three forecasting teams: one from his own Meteorological Office, one from the Admiralty (which included an Aucklander, Lawrence Hogben) and one from the United States Army Air Forces (which included another New Zealander, Jim Austin, from Otago).

The advice of these groups was often diametrically opposed. The American team used an analogue method, comparing the current weather map with maps from the past, and were often over-optimistic. The Meteorological Office, aided by the brilliant Norwegian theoretician Sverre Petterssen, had a more dynamic approach using wind and temperature observations from high altitude provided by the RAF, and were closer to the mark.

The decision to invade on Tuesday 6 June, taken late on Sunday night and finally confirmed early Monday morning, was based on a forecast of a short period of improved weather caused by a strengthening ridge following the front that had brought Monday's rain and strong winds. In the event, Monday's bad weather had already given the Allies a crucial advantage; it had put the Germans off guard (fig. 13.3).

The Germans were uncertain exactly when and where the invasion would come, but believed the most likely place was Calais and that the most likely time was July. Hitler had long understood that the key to anticipating the timing of the invasion would be good weather forecasting.

But by the summer of 1944, German weather forecasters in France were hampered by a lack of weather observations over the Atlantic because their submarine fleet was now much depleted and the Luftwaffe had largely yielded the skies to the RAF. Consequently, their forecasters could not detect the subtle changes that would lead to a temporary improvement in the conditions starting Monday evening.

Rommel, the general commanding the German defence, had identified the period 5–7 June as high risk because of the state of the moon and tide. However, he also believed the Allies would not attempt an invasion without a guarantee of six days fine weather. Reassured by a Luftwaffe weather forecaster that the bad weather starting on 5 June would last at least three days, Rommel left France for Berlin.

Consequently, he was in Germany when the invasion began and only made it back to the front at the end of the first day after being wounded en route. The German navy also dropped their guard when the bad weather commenced, and did not patrol the Channel. Only five weeks before, some of their torpedo boats had crossed the Channel and attacked a night-time rehearsal for the landings. In 10 minutes, they sank two landing craft, crippled a third, and killed over 600 sailors and soldiers.

But on the Monday night when the invasion fleet of over 6000 ships crossed the Channel, the torpedo boats did not venture out until 4.00 a.m., and the fleet had already been anchored about 15 km off the beaches along a front of 100 km for more than an hour.

The weather on 6 June was tolerable but not ideal. Strong winds scattered the paratroops, some of whom overshot the Cherbourg Peninsula and landed in the sea and were drowned. However, the Germans were also obliged to scatter their defences in response.

On the run into Omaha beach, large waves swamped 27 out of 32 amphibious tanks, and all the artillery was lost. It was here that the Allies suffered their greatest losses of the day and briefly considered withdrawing. At the end of the first day, Allied casualties were 12,000 killed, wounded and missing, as against an estimated 75,000 if surprise had not been achieved.

The weather that northern summer was among the worst on record. Several days after the landings, a storm wrecked one of the artificial harbours that had been built and caused four times the losses in ships and equipment than had occurred during the landing. Two weeks later, in the second time slot suitable for the invasion, another major storm came through prompting Eisenhower to send Stagg a letter saying, "I thank the Gods of war we went when we did".

MODERN WEATHER FORECASTING

Data Gathering Modern weather forecasting depends on as detailed a knowledge of current conditions as can be obtained. In New Zealand, this involves hourly weather observations from about 100 locations – including offshore islands like Raoul Island in the Kermadecs, the Chatham Islands, and Enderby and Campbell Islands in the southern ocean.

Many of these are provided by machines that automatically measure wind strength and direction, temperature and humidity, air pressure, rainfall in the last hour and the last ten minutes, and sometimes also lightning strikes, visibility, cloud amount and the height of the cloud base. Reports provided by people include most of the above as well as whether rain or clouds can be seen in the distance.

Satellite photos covering the Tasman Sea and New Zealand are received almost hourly, day and night. These come in two sorts: infrared, which 'see' heat radiation (fig. 11.4) emitted by the clouds, land and sea surface; and visible photos, which 'see' reflected light in the manner of a normal camera (fig 13.4). Infrared images have the advantage of being available at night. They can distinguish areas of cloud 5 km or more in diameter and give cloud-top temperatures, from which the height of the cloud top can then be calculated. During the day, visible pictures, available every hour, can distinguish clouds as small as 1 km in diameter.

On some satellites other instruments are able to look down through the atmosphere and measure the air temperature at different heights – something that could only be done by weather balloons up until about a decade ago. This is of particular importance to forecasting in the Southern Hemisphere where there is considerably more ocean than land and traditional data gathering methods are unavailable, aside from a handful of reports from ships and a few drifting buoys.

Fig 13.4. Satellite photos are a key tool of weather forecasters. This 'visible' satellite image, taken at 5.00 p.m., 6 November 1993, shows a line of thunderstorms moving east across the North Island while a narrow band of frontal cloud crosses the South Island and hooks around a low over the Tasman Sea.

Weather radars continually scan the skies around Auckland, Wellington and Christchurch producing new images every 15 minutes. These not only show where the raindrops are, but can also measure how fast a raindrop is travelling, and so give a measure of wind speed.

Weather balloons are released from eight places around New Zealand every day at midnight, 6.00 a.m. and noon. These are tracked with radar in order to find the wind speed and direction at different heights through the atmosphere. At three of these places – Invercargill, Paraparaumu and Whenuapai – the balloons released at midnight and midday carry a small box of instruments that measure temperature and humidity and transmit these back to the ground. Frequent measurements of wind and temperature at high altitude are also provided by some commercial aircraft.

Gathering weather information often involves the very latest technologies. For example, there is now a satellite that can measure wind speed over the sea by transmitting a pulse of microwave energy which is reflected back to the satellite from the sea surface. The stronger the return pulse, the larger the capillary waves on the sea surface and therefore the stronger the surface wind.

Number Crunching The great flood of weather data serves two purposes. Firstly, much of it goes into the global computer models that run several times a day producing weather maps five days in advance, and secondly, it is used to monitor and fine tune existing forecasts.

To this task the professional forecaster brings a wealth of knowledge and experience. Forecasting is not just a matter of moving the existing weather around, but also correctly anticipating the sometimes rapid growth and decay of depressions, fronts, thunderstorms and other cloud bands. Where the computer forecast maps show patterns of lines weaving over the country, the trained forecaster must visualise how the patterns of clouds, rain and sunshine will be influenced by the lie of the land.

Without access to all this data, there might seem little place for an amateur. There are, however, still gaps of hundreds of kilometres between observations where the forecaster's ideas of what is happening may be far from perfect, especially when there are cloud features too small to be properly distinguished by satellite pictures.

Forecasting with any success depends on the fact that the weather is organised to some degree on a very large scale. Referred to as the synoptic scale by meteorologists, this is the scale of a low pressure system or anticyclone and may cover as much as a million square kilometres. But while it might be possible to generalise about much of the weather under, say, the eastern or western half of an anticyclone, there are often smaller scale weather systems embedded within its area of influence that can still cover thousands of square kilometres.

Towering cumulus is a sign that showers are developing.

These are known to meteorologists as meso-scale systems. Examples are a narrow line of heavy showers embedded within an area of light showers or an area of 65 km/h (35 knot) gales tens of kilometres wide and hundreds of kilometres long within a surrounding area of winds of only half this speed (fig. 9.2). These frequently occur when there is a mountainous landmass like New Zealand interacting with the wind flow (fig.10.1).

Smaller again than meso-scale systems are micro-scale features that may only affect tens of square kilometres. Examples include the gust front caused by a downburst from a cumulus shower cloud (Chapter 7) or the way the wind speed over a ridge may be more than double that of the air upstream from the mountains.

These small-scale features can sometimes slip through the observing network undetected. Illegal rain – as it is sometimes ruefully called by forecasters – has a way of turning up from time to time.

More than this, the amount of information a forecaster has to try to cram into a forecast for the whole of New Zealand is sometimes a challenge. The many ranges of hills and mountains chop New Zealand's weather into dozens of different microclimates. Doing justice to them all in the few words allowed in one short forecast can be like trying to juggle too many balls at once.

DOING IT YOURSELF

The key way to improve on professional forecasts is through local knowledge gained from years of experience of the weather in the place you live. Observe which wind directions are associated with rain and become familiar with the various typical sequences of weather and their characteristic cloud signatures.

Particular types of cloud can foretell opposite kinds of weather depending on their location, so it's best to be cautious when taking your weather eye to different places around the country.

For example, lenticular wave clouds forming north of Mount Ruapehu in a southerly flow are generally followed by fine weather, whereas lenticular clouds forming east of the main ranges in a northwest flow are generally a sign that a front is approaching with bad weather for the ranges, and possibly elsewhere (fig. 13.5).

Fig 13.5. Lenticular cloud over the Rangipo Desert in Tongariro National Park.

One of the most well-known weather sayings relating the look of the sky to a forecast is "Red sky at night, shepherd's delight; red sky in the morning, shepherd's warning". This has some basis in fact. For a high cloud sheet to be brilliantly lit up by the setting sun, the sky to the west must be clear in order for the sun's rays to reach the high cloud. Much of our weather comes from the west, so clear skies in that direction often herald a period of fine weather.

Conversely, for the rising sun to light up high clouds, skies must be clear to the east; but the fact that there are high clouds overhead to be lit up may mean a major weather system like a front is advancing from the west.

Like most forecasting rules, this one has exceptions that can be as important as the rule itself. For example, in east coast districts rain comes more often from the south than the west. Therefore a delight-inducing display of red clouds at sunset in a westerly flow could nevertheless be followed the next day by a cold southerly change and rain (fig. 13.6).

If a cloud passes in front of the sun on a sunny day, few people can resist the urge to look up and see if rain is coming. In general, a steady increase in the amount of cloud is often a sign that rain is on the way. And the darker the cloud base, the thicker the cloud – a further sign that rain is more likely.

Fig 13.6. Red sky at night during a westerly may herald a cold southerly and rain.

The appearance of cirrus invading a clear sky (fig. 13.7) is sometimes taken as a sign that a front is coming and will bring rain within 24 hours. But this does not always work out. If a large, intense anticyclone becomes slow moving over or just east of New Zealand, as frequently happens in summer, a sequence of fronts will often move into its high pressure zone from the west over a period of many days. As each one approaches, it is subjected to the anticyclone's downward air motion, which dissipates most of the cloud, though some high level cirrus may subsequently move over the country without producing any rain.

Even those fronts in a westerly airstream which *do* produce rain, may only do so over and west of the main divide, so that eastern areas like Christchurch receive no rain, despite a display of high cloud. This was the case in Chapter 4 (fig. 4.10). On these

Where to find good weather forecasts

RADIO

Most radio stations broadcast regional forecasts hourly after the news, and marine forecasts where applicable. **National Radio** broadcasts daily MetService coastal forecasts at 3.03 a.m. and 5.03 a.m., and daily MetService mountain forecasts at 4.06 p.m.

TELEVISION

TV 1 and TV 3 broadcast comprehensive general forecasts with situation maps, satellite images and extended outlooks at the end of their 6.00 p.m. news bulletins.

NEWSPAPERS

Major daily newspapers publish situation maps and forecasts, and the best devote up to half a page to weather information with good graphics showing wind strength and direction, likelihood of rain, temperature, and extended forecasts. *The Press* Internet site (http://www.press.co.nz) has a useful weather page including a situation map and extended forecasts for all New Zealand, and links to other weather sites.

METSERVICE (Meteorological Service of New Zealand Ltd)

MetService provides to the public a wide range of forecasting services, including latest regional, marine, recreational, farming, ski and mountain forecasts, situation and forecast weather maps, and satellite images. These are supplied to the public through their Metphone and Metfax 0900 services (which are charged to your phone account), and on the Internet.

MetPhone An 0900 service offering recorded regional, rural, marine, coastal, mountain and ski forecasts, and weather outlooks for all parts of New Zealand, including national parks and other mountain areas. A list of direct dial numbers for forecasts for specific areas is available. Call Freephone 0800 500 669, or consult your telephone book, or the MetService website (see below). Regional forecasts and ski information: 0900 999 + Telecom area code.

MetFax An 0900 fax service supplying regional, coastal, mountain and rural forecasts, outlooks for all parts of New Zealand, and situation and forecast weather maps and satellite pictures. The service aims to deliver the fax within ten minutes of your call.
MetFax Helpline 0800 WEATHER (0800 932 843) for the direct dial number required for the area you are interested in, or consult the MetService website (see below).

MetService web site: http://www.met.co.nz. On this website Internet users have free access to short regional forecasts, an extended five day forecast, brief mountain forecasts, hazardous weather warnings, coastal marine and recreational marine forecasts, and a high seas forecast. Full details of the MetFax and MetPhone 0900 services are supplied, along with information about other products MetService provide via the Internet. Its 'Weather Now' package is available to subscribers to Xtra.

WEATHER ON THE WEB

Useful sources of New Zealand weather information on the Internet.
MetService: **http://www.met.co.nz** (see above)
The Press: **http://www.press.co.nz** (see above)
Victoria University: **http://www.rses.vuw.ac.nz/meteorology/index.html**
National Institute of Water and Atmospheric Research (NIWA): **http//www.niwa.cri.nz**
University of Canterbury Geography Department: **http://www.geog.canterbury.ac.nz**
Australian Bureau of Meteorology: **http://www.bom.gov.au**

occasions, people on the West Coast, where rain most certainly does fall, may not be able to observe the steady increase and lowering of high cloud because a blanket of low cloud has come in first and obscured the rest of the sky.

Of course, not all increases of cloud herald rain. In the winter half of the year a continuous layer of low cloud will often spread over Christchurch and other east coast areas in a northeast flow. Drizzle may occur and conditions can be pretty cold and miserable, but that's usually as bad as it gets – unless a depression approaches from the north

Fig. 13.7. Cirrus ahead of frontal cloud in the Mackenzie Basin.

with an easterly storm ahead of it. In order to get more than drizzle, upward air motion is required through a significant portion of the atmosphere. But much of the cloud in winter is associated with anticyclones, which cause downward air motion above the cloud layer, so no rain is produced.

If weather forecasts go wrong, it's often the timing that's a bit out, while the sequence of events will still be correct. So if you feel the weather isn't behaving exactly as predicted, try and keep the wider pattern in view. For example, there was a test match between the Lions and the All Blacks in Wellington in 1983 when a strong but brief southerly change was forecast. However, the southerly came in early, before the match had started. The All Blacks' captain, Andy Dalton, who had won the toss, chose to play with the wind in the first half knowing that it was likely to die down in the second. This in fact happened, helping the All Blacks to victory.

And of course you can always adjust your plans to take advantage of the current weather situation, like the football team on the West Coast whose home ground was up in the hills and therefore often in the clouds. They were not above slipping an extra player into the mist when the game was close!

Conclusion

We often think of the sky starting somewhere above us and dropping the weather down on our heads, but I only have to step outdoors into a good Wellington gale to be reminded that the sky starts about where the grass ends, down around ankle height.

In this sense, we are sky people. We live surrounded by the sky and we cushion ourselves from its effects with our clothes and houses. By forecasting the weather we try to sidestep the sky's worst behaviour and get out of the way of its dangerous storms as best we can.

Taking the long term view, the cycles of the sky's weather over the last two and a half million years – known as the ice ages – have shaped the land we walk on and may even have shaped us: some scientists believe that the cycling in and out of glacial conditions has played a key role in accelerating human evolution.

Taking a short term view, it has been estimated that water molecules in a falling raindrop will spend 10,000 years in the sea after only nine days in the sky. Of course, some rainwater evaporates quickly and sneaks a second trip to the sky, but snow falling on a polar ice cap may be locked up in the ice for hundreds of thousands of years.

As we explore the sky and learn more about it, we are challenged by the knowledge that the atmosphere is finite and we humans are capable of changing it in ways that threaten our lives. High levels of pollution in cities can cause lung disease; chemicals produced for hairspray and fridges have attacked the ozone layer, threatening to allow dangerous ultraviolet radiation to reach the Earth's surface; increases we have made to the greenhouse gases by burning fossil fuels may change the Earth's climate.

Lenticular wave cloud, at sunset, over Tasman Bay.

Faced with these threats we must take a deep breath and accept another challenge: to step into the role of guardians of the sky and weather, as well as the rest of life on Earth.

In writing this book I have tried to explain how the weather works, not only to help readers avoid its dangers, but also to share the excitement I feel as a weather forecaster anticipating the unfolding drama of each day's weather. If I have helped you to enjoy the weather, then I have succeeded.

CONTRIBUTORS

Weather maps were reproduced by One Sky Design from originals produced by the Meteorological Service of New Zealand Ltd and its predecessor organisation. Map for D-Day sourced from *Forecast For Overlord*, by J M Stagg, published by Ian Allan Ltd, UK, 1971.

Diagrams and graphs were drawn by Jo Williams and Leon Dalziel of Nimbus Advertising, except for: figures 4.4, 6.4, and 7.8 (courtesy *New Zealand Geographic* & Glenn Conroy); figure 7.3 (originally painted by Dave Gunson for *New Zealand Geographic*); figure 11.3 (MetService); figure 12.1 (New Zealand Meteorological Service); figures 12.2–12.4 (NIWA); figure 12.5 (Bureau of Meteorology, Australia); figures 12.7, 12.10 and 12.11 (sourced from *Climate Change – The IPCC Scientific Assessment* published by World Meteorology Association/UNEP, 1990).

The engraving 'Recent Bushfires' on page 38 was supplied by the Alexander Turnbull Library (C- 19032). The sketch of Kaiapohia Pa (fig. 13.1) is from *The Stirring Times of Te Rauparaha*, by W T L Travers, Whitcombe & Tombs Ltd, n.d.

Satellite photos: Japan Meteorological Agency.

Radar maps in Chapter 10 courtesy of Warren Gray, NIWA.

Photographs: **Cover** & author photograph: Craig Potton. Remaining photos by Craig Potton except for the following: **32** Dave Wethey; **36** Dionne Ward, Evening Standard; **38** Serials Collection, Alexander Turnbull Library (C 19032); **41** Don Scott, *The Press*; **49** Wellington Maritime Museum Collection (N676); **52** Wilson and Horton Ltd; **54** David Chowdhury; **57** *NZ Herald*; **61** Glen Fergusson, *Daily News*; **67** Wayne Carran, Works Civil Construction; **76** Making New Zealand Collection, Alexander Turnbull Library (F 360 1/4); **82** *Evening Post* (4652-1976); **91** *NZ Herald*; **92** National Newspaper Collection, Alexander Turnbull Library (N-P 331); **93** (top) National Newspaper Collection, Alexander Turnbull Library (N-P 332), (bottom) *Gisborne Herald* (D7895); **95** *Evening Post* (1653-1968); **110** (top) Arno Gasteiger; (bottom) Quentin Christie/NZ Society of Soil Science.

FURTHER READING

Among the scores of books on meteorology the following recommend themselves for clarity, New Zealand content, or entertainment value.

Ahrens, C D 1994, *Meteorology Today*, 5th edn, West Publishing Company, Minnesota. A comprehensive but accessible textbook. Copious colour illustrations. Focused on United States and Northern Hemisphere so some diagrams need to be held upside down in front of a mirror.

Bohren, C F 1987, *Clouds in a Glass of Beer – Simple Experiments in Atmospheric Physics*, John Wiley & Sons, New York. Excellent collection of essays on atmospheric phenomena with many interesting experiments.

Crowder, R 1995, *The Wonders of the Weather*, Australian Government Publishing Service, Canberra. Well written, accessible, excellent illustrations, aimed for general public although focused on Australia.

Greenler, R 1980, *Rainbows, Halos, and Glories*, Cambridge University Press, Cambridge, Mass. An excellent well-illustrated book on atmospheric optical phenomena.

Holford, I 1977, *The Guinness Book of Weather Facts and Feats*, Guinness Superlatives Ltd, Enfield, UK. Wide ranging, informative, easy to dip into, well illustrated.

Junger, S 1997, *The Perfect Storm*, Fourth Estate, London. Stunning account of the fatal impact of explosive cyclogenesis on fishing boats off the east coast of the United States.

Lamb, H H 1982, *Climate, History and the Modern World*, Methuen, London. One of the leading climatologists of the 20th century describes in detail the impact of climate change on human civilisation throughout the last 10,000 years.

McGlone, M, Clarkson, T, & Fitzharris, B 1990, *Unsettled Outlook – New Zealand in a Greenhouse World*, GP Books, Wellington. A detailed consideration of the effects of climate change and depletion of the ozone layer on New Zealand.

Sturman, A & Tapper, N 1996, *The Weather and Climate of Australia and New Zealand*, Oxford University Press, Melbourne. A good textbook with a local flavour. Southern Hemisphere diagrams which do not need to be held upside down in front of a mirror.

INDEX